THE NEW DEMOCRACY WARS

The International Political Economy of New Regionalisms Series

The International Political Economy of New Regionalisms Series presents innovative analyses of a range of novel regional relations and institutions. Going beyond established, formal, interstate economic organizations, this essential series provides informed interdisciplinary and international research and debate about myriad heterogeneous intermediate level interactions.

Reflective of its cosmopolitan and creative orientation, this series is developed by an international editorial team of established and emerging scholars in both the South and North. It reinforces ongoing networks of analysts in both academia and think-tanks as well as international agencies concerned with micro-, meso- and macro-level regionalisms.

The New Democracy Wars
The Politics of North American Democracy Promotion in the Americas

NEIL A. BURRON

ASHGATE

Published by
Ashgate Publishing Limited
Wey Court East
Union Road
Farnham
Surrey, GU9 7PT
England

Ashgate Publishing Company
110 Cherry Street
Suite 3-1
Burlington, VT 05401-3818
USA

www.ashgate.com

British Library Cataloguing in Publication Data
Burron, Neil A.
    The new democracy wars : the politics of North American
    democracy promotion in the
    Americas. -- (The international
    political economy of new regionalisms series)
    1. Democratization--Government policy--United States.
    2. Democratization--Government policy--Canada.
    3. Democratization--Bolivia. 4. Democratization--Peru.
    5. Democratization--Haiti. 6. United States—Foreign
    relations--21st century. 7. Canada--Foreign relations—
    21st century. 8. Hegemony--Case studies.
    I. Title II. Series
    327.7–dc23

Library of Congress Cataloging-in-Publication Data
Burron, Neil A.
    The new democracy wars : the politics of North American democracy promotion in the Americas /
by Neil A. Burron.
        p. cm. -- (The international political economy of new regionalisms series)
        Includes bibliographical references and index.
        ISBN 978-1-4094-4906-5 (hardback) -- ISBN 978-1-4094-4907-2 (ebook)
    1. Democratization--Government policy--United States. 2. Democratization--Government policy--
Canada. 3. United States--Foreign relations--Latin America. 4. Latin America--Foreign relations--United
States. 5. Canada--Foreign relations--Latin America. 6. Latin America--Foreign relations--Canada. I. Title.
    JZ1480.A53B87 2012
    327.1'1--dc23

        2012018437

ISBN 9781409449065 (hbk)
ISBN 9781409449072 (ebk – PDF)
ISBN 9781409471875 (ebk – ePUB)

Printed and bound in Great Britain by the
MPG Books Group, UK.

Contents

For my girls, Kelly-Anne and Ella.

List of Illustrations

Illustration

List of Tables

Foreword

'The struggle over the meaning of democracy – and who possesses the legitimacy to promote it – constitutes the political backdrop of the investigation in the pages that follow.' So writes Neil Burron near the beginning of his first chapter. Indeed, analysis of the meanings that lie behind words used by authors and politicians – most particularly highly politicized words like democracy – is the first responsibility of the critical writer. The next responsibility is to place the discussion of political meanings in the context of the power relationships surrounding the actions or institutions under consideration. Words are tools of power. The critical author's job is to make clear who, in their use of words, is wielding power and to what end.

This book is written in that vein. It is an analysis of the evolution of power relationships in the Americas during the late twentieth and early twenty-first centuries. Such an analysis is to be read as a regional focus on a general process of restructuring world power; and in this task the common vocabulary is often misleading, a vocabulary inherited from the seventeenth century establishment of the modern Westphalian state system in Europe. Then the theory was that the 'state' was independent in relation to other states and that it had a monopoly of authority within its borders. Now we know that neither of these conditions holds fully, although, as between more powerful and weaker states, there is a question of degree as to what extent it does hold.

In the late twentieth century the word 'globalization' became current, particularly among people who championed the idea that global capitalism, expanding with freedom to work its way through compliant state structures, was binding the world into an integrated capitalist whole; and that this, so its champions thought, would be for the benefit of all mankind by creating continuous economic growth.

Others did not see things in the same light. They saw a sequence of financial upheavals and a steadily growing disparity in incomes throughout the world. This historical trend became caricatured as financial managers were paid colossal bonuses while their failing banks and financial institutions were saved from collapse by bailouts from the public purse.

The current global disparity in political economic power relations can be traced back to the Asian financial crisis of 1997–1998. It was the first major warning signal since the new model neoliberal world economy set up in the 1970s that all was not right with the system; but the victim countries were at the time relatively poor and powerless. The crisis of 1997 left the weaker Asian economies open to plunder by foreign capital. A Japanese proposal to create a regional monetary fund as a means toward sustaining struggling Asian economies was rejected by the United States and the International Monetary Fund. The neoliberal order prevailed but the Asian countries remembered, took thought, and began to devise policy to avoid a 'next time'.

There has been a slow but steady movement away from consensus about or acquiescence in the neoliberal world order. The Asian financial crisis of 1997–1998, which was seen by many in the West as the failure of the East Asian model of state-led economic growth, was seen in Asia as a demonstration of the dangers in the way global financial matters were managed out of Washington and Wall Street. Its lesson was the need to reduce the vulnerability of Asian countries and their societies to global financial markets and to the IMF which had come down hard upon them. In retrospect, that Asian crisis of 1997–1998 may have been the turning point away from a worldwide consensus on the neoliberal order. Latin American countries echoed the Asian experience; and in some of these countries internal revolts radically contested US hegemony.

Since the Asian crisis, massive US trade and budget deficits have been financed principally by China and Japan; and on top of this American public and private indebtedness, which strained confidence in the dollar, came the sub-prime mortgage binge that triggered the collapse of financial markets in 2008 and the prospect of a prolonged worldwide recession.

This time governments, beginning with the US government, felt obliged to move more forcefully into the economy to repair the damage done by unregulated capitalism. The erstwhile dominant economies of the West had to adjust to a new global balance of economic power. As the major capitalist countries in North America and Europe became engulfed in economic crisis, the Asian countries and some others in the South, including Brazil in South America, continued to grow.

The United States was manifesting more and more an imbalance in its political economy as between its military and economic security. Since the de facto defeat of their armed forces in Vietnam, government and people of the United States had quietly turned their backs in shame on the civilian conscript army. A new volunteer professional army was built up by the Pentagon into a quasi-independent center of power within the state, with over 700 bases outside US territory to assure the deployment of its 'full spectrum dominance' over Asia, the Middle East, Africa and Latin America as well as its influence in the countries of North America and Europe.

The United States had become the sole military superpower; but its military power was sustained by an economy that was weakening relative to the emerging economic great powers, China and India in Asia, Russia in Eurasia and Brazil in Latin America – the BRICS. This global imbalance in military/economic power which has emerged during the first decade of the twenty-first century contains constantly the risk of being shattered by some rash military action – an action such as an attack by Israel on Iran's nuclear facilities – the consequences of which would be unpredictable but potentially catastrophic for all.

This book focuses on the changes in power relationships in North and South America that have been taking place within the very delicate process of global rebalancing of power still going on and by no means near completion. We see here in the pages that follow the effort of US foreign policy aided and abetted by US-directed civil society to sustain and strengthen neoliberal structures of power within the more amenable Latin American countries. We see also the revolt of popular (or 'populist' if you don't like them) forces in some Latin American countries and the emergence of new patterns of sub-regional alliances among those countries in which the balance of domestic political power has shifted from elites managing neoliberal policies to movements of reform supported by organized strength among the relatively disadvantaged sectors of society.

We see also a consolidation of North American power in an alignment of Canada with the weakening influence of the United States in Latin America. Not so long ago Canadians thought in an East-West orientation towards Europe on one side and Asia on the other. This orientation, firmly implanted in the Canadian population since the American Revolution, was strengthened by the growing numbers of Southern and Eastern Europeans and of South and East Asians who had joined in Canada's heterogeneous population more recently.

Canadian foreign policy post-World War II was independent of the United States foreign policy, notably in the initiative taken by Lester Pearson's government to make the insertion of Canadian peace-keeping military forces between combatants in situations of potentially violent conflict a major element of Canadian foreign policy. Pearson's conception of Canada's role in the world was often characterized as 'middlepowermanship.' The meaning of the term was explained by that venerable theorist and practitioner of Canadian foreign policy, John Holmes, not as Canada being a middle level power, but as Canada being the power 'in the middle' as an independent mediator between conflicting forces.

This East-West orientation of Canadians' thinking, which embraced the whole world, gradually, however, gave place during the late twentieth century to a stronger North-South orientation connecting Canada both economically and culturally with her southern neighbour. During the first decade of the twenty-first century this connection has become increasingly political.

The shift was marked significantly by the rest of the world when Canada lost its seat on the United Nations Security Council, showing that Canada's reputation for impartiality and independence in the world community had evaporated as Canada became instead regarded as invariably aligned with the United States and its hegemonic aims. Canada, once in a mediating position in the Arab/Israeli conflict became with the Harper government unconditionally identified with Israel. This is to say that, at the global level, Canada forfeited its once widely respected independence to become perceived by the rest of the world as wholly committed to the United States just at a time when US world leadership was weakening and new political/economic forces were making for a more plural world order.

Canada's role in the Western Hemisphere fits into this picture. Neil Burron shows in this volume how democracy promotion has evolved in Canadian policy to conform with the American approach to democracy promotion. So Canada's democracy promotion in Latin America now serves a hegemonic role, helping to reduce the scope of democracy in target Latin American countries to the requirements of neoliberal accumulation.

The grass roots project of promoting democratic arousal and making it a force for comprehensive social change, which has animated some Canadian non-governmental organizations in the international and regional fields, is being increasingly excluded from state sponsored aid programs. If anything like the former independence of Canada's stand in the world, as in the hemisphere, is to be salvaged, it would be through a renewed assertion of support in Canada's civil society for the social and political mobilization of the disadvantaged sectors of Latin American societies. This would imply political alignment of Canadian civil society activism with the emergence of regional groupings of the more leftist Latin American countries, a prospect hedged with some doubt as to its likelihood.

Suddenly, however, challenge to the neoliberal hegemony emerged during 2011 within the United States itself in the non-violent, frequently ironic, Occupy Wall Street movement that began where its name suggests and then spread throughout the United States and beyond mobilizing citizens to demonstrate in streets and in parks. Ignored initially by the mainstream media as a dismissible phenomenon, it gradually became an acknowledged challenge to the neoliberal order at its very source. It has achieved no substantial change, indeed may have stimulated a tightening of the hegemonic order, but it has raised consciousness to belief that another world order is conceivable.

Robert W. Cox

Preface

The idea for this book began following my experience working in Haiti during the winter of 2005–2006, where I served as a coordinator for a Canadian-led international electoral observation mission in Port-au-Prince. The elections in question were meant to restore democracy to the beleaguered Caribbean republic after another coup led to a break in the constitutional order in February 2004. The encounter exposed me to the contradictory reality of Canada and the international community's attempt to re-establish democracy while undermining it in so many other ways, and compelled me to explore this paradox further. Although the impetus to write on this topic was deeply personal, the book would not have been possible without the material and intellectual support of several others.

The field research carried out in Haiti, Bolivia and Peru was made possible by a Canadian Window on International Development grant from the International Development Research Centre. In the course of conducting the field research, I had the privilege of meeting many individuals whose commitment to social change was both an inspiration and a call to action. I thank them for generously sharing their time and knowledge with a foreigner. I also met with many Canadian and US officials in the field, as well as senior officials in Ottawa. Although this book is critical of many of the policies they implemented, I thank them for the candidness with which they typically spoke.

A well-established critical literature on US democracy promotion provided the initial bearings for my investigation, and this book owes a considerable debt to the work of William Robinson and Adam David Morton in particular. Along with Robert Cox's seminal work on world order, their work provided the launching pad for the comparative critical account of Canadian and US approaches to democracy promotion in the pages that follow. Laura Macdonald, Cristina Rojas and Randall Germain at the Department of Political Science at Carleton University were instrumental in helping me sharpen my arguments and situate them within the vast literature on democracy promotion, democratization, and international political economy. This book would not have come to fruition without their guidance. Professor Yasmin Shamsie of Wilfred Laurier University provided invaluable feedback on my analysis of Haiti, and Professor Stephen Clarkson of the University of Toronto encouraged me to begin publishing my material after reading an initial paper on Canadian and US approaches to democracy promotion presented at the annual meeting of the Canadian Political Science Association in June 2010.

Timothy Shaw, Series Editor, believed in this project from the outset, and provided many helpful suggestions to situate the analysis within the larger developments characterizing hemispheric relations. His knowledge and expertise were greatly appreciated. Kirstin Howgate and Nikki Selmes provided helpful and detailed editorial guidance and graciously responded to my many questions on style and format.

Some of the arguments made in this book have appeared previously in article form. Publisher's permission to draw material from the following articles was gratefully received: 'Unpacking U.S. Democracy Promotion in Bolivia', *Latin American Perspectives*, 39 (1) (2012), 115–132; 'Curbing 'Anti-Systemic' Tendencies in Peru', *Third World Quarterly*, 32 (9) (2011), 1655–1672; 'Reconfiguring Canadian Democracy Promotion – Convergence with the U.S Approach?', *International Journal*, 66 (2, Spring) (2010), 391–417; 'Ollanta Humala and the Peruvian Conjuncture: Democratic Expansion or 'Inclusive' Neoliberal Redux?', *Latin American Perspectives*, 39 (1) (2012), 133–139; 'No Smoking Gun – Yet: Canadian Democracy Promotion in Bolivia', *NACLA*, 43 (3, May) (2010). The arguments developed in this book benefitted greatly from the reviewers who provided feedback on the manuscripts that were submitted for publication. The content from these articles has been updated and revised, however, to produce a new cohesive, integrated account.

Family and friends provided the moral support without which I might have decided long ago to abandon this project. The encouragement and support of my parents, who have always nurtured my curiosity, was particularly important. Two friends in particular were instrumental in providing comments on early drafts of the opening chapter – Nathaniel Grimkin and José Silva. Above all, my wife and life partner, Kelly-Anne, provided the spiritual encouragement, support and understanding which made the writing of this book possible. As my research-travel companion, moreover, many of the ideas presented here were developed and sharpened through our many exchanges throughout the interviewing process. Above all, I appreciate her constant reminder to 'remain objective' when assessing deeply subjective issues. Before I could begin publishing the results of my research, we were blessed with a beautiful girl, Ella, whose presence has enriched our lives in so many ways it is hard to remember what things were like without her. I am grateful to Kelly-Anne for taking on more than her fair share of childrearing responsibilities and to Ella for always welcoming me back into the world of spirited discovery when it was finally time to turn off the computer. This book is dedicated to both of them.

Neil A. Burron

About the Author

Neil Burron completed a PhD in political science at Carleton University. He has worked in the field of international development with various NGOs and has travelled widely in Latin America. Dr. Burron is currently working as an independent researcher in the Ottawa region

List of Abbreviations

ACILS	American Center for International Labor Solidarity
ACL	Asociación Civil Labor (Civil Labour Association)
ACOBOL	Asociación de Concejalas de Bolivia (Association of Women Councillors of Bolivia)
ADEX	Asociación de Exportadores (Association of Exporters)
ADN	Acción Democrática Nacionalista (Democratic Nationalist Action)
AIDESEP	Asociación Interétnica de Desarrollo de la Selva Peruana (Inter-ethnic Association for the Development of the Peruvian Jungle)
ALBA	Alianza Bolivariana para los Pueblos de Nuestra América (Bolivarian Alternative for Latin America)
ANMH	Association National des Medias Haïtiens (National Association of Haitian Media)
APRA	Alianza Popular Revolucionaria Americana (American Popular Revolutionary Alliance)
APRODEH	Asociación Pro Derechos Humanos (Pro-Human Rights Association)
BAI	Bureau des Avocats Internationaux (Bureau of International Lawyers)
BRIC	Brazil, Russia, Indian and China
CAJ	Comisión Andina de Juristas (Andean Commission of Jurists)
CARICOM	Caribbean Community and Common Market
CARLI	Comité des Avocats pour le Respect des Libertés Individuelles (Committee of Lawyers for the Respect of Individual Rights)
CCAD	Canadian Centre for the Advancement of Democracy
CCI	Cadre de Coopération Intérimaire (Interim Cooperation Framework)
CCP	Confederación Campesina del Peru (Confederation of Peasants of Peru)
CCPSC	Comité Cívico Pro Santa Cruz (Civic Committee of Santa Cruz)
CEADESC	Centro de Estudios Aplicados a los Derechos Económicos Sociales y Culturales (Centre of Applied Studies for Economic, Social and Cultural Rights)
CEAS	Comisión Episcopal de Acción Social (Episcopal Commission for Social Action)
CEDEP	Centro de Estudios para el Desarrollo y la Participación (Centre of Studies for Development and Participation)

CELAC	Comunidad de Estados Latinoamericanos y Caribeños (Community of Latin American and Caribbean States)
CEP	Conseil Electoral Provisoire (Provisional Electoral Council)
CEPB	Confederación de Empresarios Privados de Bolivia (Confederation of Private Businessmen of Bolivia)
CGTP	Confederación General de Trabajadores del Perú (General Confederation of Workers of Peru)
CHIRAPAQ	Centro de Culturas Indígenas del Perú (Centre of Indigenous Cultures of Peru)
CHPP	Conférence Haïtienne des Partis Politiques (Haitian Conference of Political Parties)
CIDA	Canadian International Development Agency
CIDOB	Confederación de Pueblos Indígenas de Bolivia (Confederation of Indigenous Peoples of Bolivia)
CIES	Consorcio de Investigación Económica y Social (Economic and Social Research Consortium)
CIPE	Center for International Private Enterprise
CLADEM	Comité de América Latina y el Caribe para la Defensa de los Derechos de la Mujer (Latin American and Caribbean Committee for the Defence of Women's Rights)
CLED	Centre pour la Libre Entreprise et la Démocratie (Centre for Free Enterprise and Democracy)
CNDDHH	Coordinadora Nacional de Derechos Humanos (National Coordinator for Human Rights)
CNE	Corte Nacional Electoral (National Electoral Court – Bolivia)
CPI	Canadian Petroleum Institute
COB	Central Obrera Boliviana (Bolivian Labour Central)
COMIBOL	Corporación Minera de Bolivia (Mining Corporation of Bolivia)
COFADEH	Comité de Familiares de Detenidos Desaparecidos en Honduras (Committee of Relatives of the Disappeared in Honduras)
CONACAMI	Confederación Nacional de Comunidades del Perú Afectadas por la Minería (National Confederation of Communities affected by Mining)
CONAP	Coordination Nationale de Plaidoyer pour les Droits des femmes (National Coordinator for the Advocacy of Women's Rights)
CONSODE	Consorcio Sociedad Democrática (Consortium for a Democratic Society)
CORECAM	Coordinadora Regional por el Cambio (Regional Coordinator for Change)
CPESC	Coordinadora de Pueblos Etnicos de Santa Cruz (Coordinator for the Ethnic Peoples of Santa Cruz)
CPH	Concertation pour Haïti (Concertation for Haiti)
CSUTCB	Confederación Sindical Única de Trabajadores Campesinos de Bolivia (Union Confederation of Peasant Workers of Bolivia)

CTH	Confédération des Travailleurs Haïtiens (Confederation of Haitian Workers)
DEA	Drug Enforcement Agency
DFAIT	Foreign Affairs and International Trade Canada (originally named Department of Foreign Affairs and International Trade)
ECLAC	Economic Commission for Latin America and the Caribbean
FAM	Federación de Asociaciones Municipales (Federation of Municipal Associations)
FARC	Fuerzas Armadas Revolucionarias de Colombia (Armed Revolutionary Forces of Colombia)
FBDM	Fundación Boliviana para la Democracia Multipartidaria (Bolivian Foundation for Multiparty Democracy)
FCM	International – Federation of Canadian Municipalities
FENATRAHOB	Federación Nacional de Trabajadoras del Hogar de Bolivia (National Federation of Domestic Workers of Bolivia)
FEUH	Fédération des Étudiants Universitaires d'Haïti (Federation of Haitian University Students)
FF	Forum of Federations
FL	Fanmi Lavalas
FLRN	Front pour la Libération et la Reconstruction Nationale (Front for National Liberation and Reconstruction)
FMOCCP	Federación de las Mujeres Organizadas en Centrales de Cocinas Populares (Federation of Women Organized in Popular Kitchens)
FNH	Fondation Nouvelle Haïti (New Founding of Haiti)
FNMCIOB	Federación Nacional de Mujeres Campesinas, Indígenas Originarias Bartolinas Sisa (National Federation of Peasant Women, Indigenous Origins Bartolinas Sisa)
FTA	Free Trade Agreement
FTAA	Free Trade Agreement of the Americas
FTQ	Fédération des travailleurs et travailleuses du Québec (Federation of workers of Quebec)
FTSMB	Federación Sindical de Trabajadores Mineros de Bolivia (Union Federation of Mining Workers of Bolivia)
FUNDAPPAC	Fundación de Apoyo al Parlamento y a la Participación Ciudadana (Foundation for Support to Parliament and Citizen Participation)
GPC	Grupo Propuesta Ciudadana (Group for a Citizen's Proposal)
GRESP	Grupo Red de Economía Solidaria del Perú (Network of Economic Solidarity of Peru)
IADC	Inter-American Democratic Charter
IDEA	Instituto de Derecho y Economia Ambiental (Institute for Rights and Environmental Economy)
IDEA	International Institute for Democracy and Electoral Assistance

IDEPUCHP	Instituto de Democracia y Derechos Humanos de la Pontificia Universidad Católica del Perú (Democracy and Human Rights Institute of the Catholic Pontificate University of Peru)
IDRC	International Development Research Centre
IFES	International Foundation for Electoral Systems
IGH	Interim Government of Haiti
IIPS	Instituto de Investigación y Capacitación Pedagógica y Social (Institute for Research and Pedagogical and Social Capacity Building)
IJDH	Institute for Justice and Democracy in Haiti
ILD	Instituto Libertad y Democracia (Institute for Liberty and Democracy)
INRA	Instituto Nacional de Reforma Agraria (National Institute for Agrarian Reform)
IOM	International Organization for Migration
IPEDEHP	Instituto Peruano de Educación en Derechos Humanos y la Paz (Peruvian Institute for Education in Human Rights and Peace)
IPTK	Instituto Politécnico Tomás Katari (Bolivian NGO)
IPYS	Instituto Prensa y Sociedad (Institute Press and Society)
IRI	International Republican Institute
LAPOP	Latin American Public Opinion Project
MAS	Movimiento al Socialismo (Movement towards Socialism)
Mercosur	Mercado Común del Sur (Common Southern Market)
MIEEH	Mission international d'évaluation des élections en Haïti (International Mission for Monitoring Haitian Elections)
MIF	Multinational Interim Force
MINUSTAH	Mission des Nations Unies pour la Stabilisation en Haiti (United Nations Mission for Stabilization in Haiti)
MIR	Movimiento de la Izquierda Revolucionaria (Revolutionary Movement of the Left)
MMR	Movimiento Manuela Ramos (Movement Manuela Ramos)
MNR	Movimiento Nacionalista Revolucionario (National Revolutionary Movement)
MPP	Mouvement Paysan Papaye (Peasant Movement of Papaye)
NAFTA	North American Free Trade Agreement
NCHR	National Coalition for Haitian Rights
NDI	National Democratic Institute for International Affairs
NED	National Endowment for Democracy
NGO	Non-governmental organization
OAS	Organization of American States
OAU	Organization of African Unity
OPs	Organisations Populaires (Popular Organizations)
OPL	Organisation du Peuple en Lutte (Organization of People in Struggle)

OSI	Open Society Institute
OTI	Office of Transition Initiatives (part of USAID)
PAJ	Programme pour une Alternative de Justice (Programme for an Alternative Justice)
PAPDA	Plate-forme Haïtienne de Plaidoyer pour un Développement Alternatif (Haitian Platform for the Advocacy of an Alternative Development)
PD	Plate-forme Démocratique (Democratic Platform)
PERCAN	Peru-Canada Mineral Resources Reform Project
PIH	Partners in Health
PJ	Red de Participación y Justicia (Network of Participation and Justice)
PNH	Police Nationale d'Haïti (Haitian National Police)
PNP	Partido Nacionalista Peruano (Peruvian Nationalist Party)
PP	Perú Posible (Peru Possible)
POHDH	Plate-forme des Organisations Haïtiennes de Droits Humains (Platform of Haitian Human Rights Organizations)
PROÉTICA	Consejo Nacional para la Ética Pública (National Council for Public Ethics)
PTPA	Peru-US Trade Promotion Agreement
QUANGO	Quasi non-governmental organization
RLCD	Red Latinoamericana y del Caribe para la Democracia (Assembly of the Latin American Network for Democracy)
RNDDH	Réseau National de Défense des Droits Humains (National Network for the Defence of Human Rights)
SCFAID	Standing Committee on Foreign Affairs and International Development
SCFAIT	Standing Committee on Foreign Affairs and International Trade
SOFA	Solidarité Femmes Haïtiennes (Haitian Women's Solidarity)
SPDA	Sociedad Peruana de Derecho Ambiental (Peruvian Society for Environmental Rights)
SPP	Security and Prosperity Partnership
START	Stabilization and Reconstruction Task Force
UJC	Unión Juvenil Cruceñista (Union of Santa Cruzan Youth)
UMPABOL	Unión de Mujeres Parlamentarias de Bolivia (Union of Women Parliamentarians of Bolivia)
UNASUR	Unión de Naciones Suramericanas (Union of South American Nations)
USAID	United States Agency for International Development
WMD	World Movement for Democracy (initiative launched by the NED)

Haitians dressed as soldiers protesting western interventionism in Jacmel, a town in southern Haiti, during carnival in 2007 (Photo: Wojciech Jakobiec).

Introduction

Few parts of the world have witnessed such extraordinary acts of self-determination as Latin America over the last decade. From the Rio Grande to the Andean cordillera, indigenous people, peasants, workers, women's groups and students have revolted against the politics of exclusion and inequality, demanding greater control over the decisions that affect their lives and the resources they produce and consume. Three decades after the first transitions from authoritarianism began – and ten years after the Inter-American Democratic Charter (IADC) inaugurated a new hemispheric democratic order – governments and social movements alike continue to struggle over the meaning of democracy and popular sovereignty. At the heart of the conflict lies the thorny relationship between capitalism and democracy and the scope of popular participation at various levels of decision making. In a region rife with democratic experiments, new left and centre-left governments have sought to balance market imperatives with the need to democratize the economy and expand the institutional space for popular involvement – often with increased executive power to lead the process of reform. Some have not gone far enough, and the movements which have served as the support base for the left governments have initiated new cycles of mobilization. In countries where conservative and centre-right governments have remained in power, foundational questions on the nature of democracy are no less settled, and popular movements continue to call for social change.

On this terrain of contention, the liberal democracies to the North have championed their own vision of hemispheric democratic development, insisting on the necessity of democratic order as the political counterpart to stability, security, development and free market prosperity. Governments of both Canada and the United States have consistently prioritized democracy in regional foreign policy, spending ever-growing amounts on bilateral and multilateral democracy assistance, promoting democratic norms through the Organization of American States (OAS) and denouncing those states which have failed to meet their standards of democratic practice.[1] At the tenth anniversary commemoration event of the IADC in Valparaíso, Chile, Canadian Minister of State Diane Ablonczy enumerated the new threats to democracy – including the concentration of

---

1 It is useful to distinguish from the outset the term democracy promotion, which refers to a wide range of political and economic strategies intended to promote democracy (including coercive ones, such as military strikes), from democracy assistance, which focuses more specifically on projects and programmes designed to support democratic development (Burnell 2008). One, of course, is a component of the other.

executive power – warning that 'backsliding and threats must be confronted or we risk having setbacks to the significant progress we have made on extending and enhancing democratic governance in the Americas' (DFAIT 2011a). Insisting on his country's adherence to the core Charter principle of non-intervention, US Deputy Secretary of State William Burns delivered his own list of democratic infractions, calling on other states to 'speak out, stand firm and act with the clarity of our convictions in defence of democratic principles' (US Department of State 2011). Yet, despite the contention surrounding the meaning of democracy – and disagreement over who or what constitutes a threat to it – much of the literature on democracy promotion has abstracted this form of engagement from the interests of its practitioners, setting it apart from foreign policy considerations linked to the larger social and political struggles characterizing Latin America and the Caribbean in the new millennium.

This book provides a critical account of US and Canadian democracy promotion in the Americas by focusing on its political dimensions. It examines the politics behind how the United States and Canada go about promoting democracy – the type of democracy each favours, the foreign policy objectives that explicitly and implicitly drive this form of engagement and the political outcomes which democracy promoters seek to achieve. It argues that democracy promotion is typically formulated to advance commercial, geopolitical and security objectives that conflict with a genuine commitment to democratic development. Defenders of democracy promotion are not without reason when they argue that democracy assistance contributes to pluralism by providing support to a range of political parties and civil society organizations of various ideological stripes (Allen 2005). US and Canadian democracy promotion agencies as a whole implement hundreds of programmes across the globe, undoubtedly providing support to actors from across the political spectrum. Yet with the shift to the Left of much of Latin America, US and Canadian democracy promotion has been used to stabilize the neoliberal state and contain the spread of left 'populism'. Both have used democracy promotion to construct a regional order in the Americas predicated on neoliberal economics and low-intensity democracy (or polyarchy),[2] and both have defined democratic backsliding selectively on the basis of political and ideological threats to this order. As the metaphor evoked in the title of this book suggests, a

---

2   The term polyarchy was coined by the political scientist Robert Dahl (1971) to signify political systems based on electoral contestation and the right of citizens to participate. It is used critically by Robinson (1996) to describe a form of democracy that does not actually involve power (*cratos*) of the people (*demos*). Rather, mass participation in decision-making is limited to choosing among elites among tightly controlled electoral processes and the political elite responds largely to the interests of capital. Wilson, Gills, and Rocamora (1993) use the term 'low-intensity' democracy in a similar vein. It should be noted that Dahl (1985) himself later warned that capitalism tends to 'produce inequalities in social and economic resources so great as to bring about severe violations of political equality and hence of the democratic process'.

war over the meaning and scope of democracy is taking place, and many North American actors have declared which side they are on.

The extent to which 'populist' states such as Bolivia, Venezuela or Ecuador have violated liberal institutions is of course a matter for debate. Those who acknowledge the political dimension of democracy promotion would argue that democracy promoters are acting in part to defend the peoples of the Americas from democratic backsliding by strengthening liberal-democratic institutions and supporting political pluralism.[3] But to accept this argument is to overlook the oligarchic nature of the state in Latin America prior to the upsurge. Populist – or popular – governments came to power precisely because the liberal-democratic states that emerged in the 1980s and 1990s continued to be governed by cliques of elites responding to the interests of dominant classes. During the democratic transitions, transnational actors – including US democracy promotion agencies – supported the efforts of moderate elites to restrict the scope of democracy as different forces struggled over the nature of popular political participation and involvement in the economy. While the rejection of political violence and increased potential for democratic participation offered by the victory of liberal democracy represented undeniable progress from the dark days of dictatorship, the adoption of neoliberal policies promoted by the United States and Canada ensured that the state continued to respond to the interests of the most powerful. The popular mobilizations against the neoliberal state have called for social transformation, and the popular governments that have come to power in their wake have responded to this aspiration. And while there is a strong case to be made that some of the new left regimes have violated liberal-democratic norms without advancing the cause of social change, such debates are secondary here. My point is that the United States and Canada have used democracy promotion to protect the neoliberal state, and that there is nothing democratic about this. How they have done this is the focus of this book.

The macro-structural backdrop to the analysis is a critique of North American-led regional order, a concept which I derive from Robert Cox's seminal work on world order to call attention to relations of wealth and power in the inter-American system (see Cox and Sinclair 1996). With the proliferation of regional projects in the hemisphere, this conceptualization traces one of the main fault lines to emerge in recent years – the growing chasm between North America and the nations of Latin America and the Caribbean. North America is thus defined in the narrow sense; although Mexico is party to the North American Free Trade Agreement (NAFTA), it is also part of the *Comunidad de Estados Latinoamericanos y Caribeños* (Community of Latin American and Caribbean States – CELAC), a new regional bloc consisting of all sovereign countries in the Americas except

---

3   The aversion to 'populism' on the part of the of George W. Bush administration, moreover, is particularly deceitful given that a significant wing of the Republican Party can be characterized as right-wing populist, and that Bush himself based his appeal on populist overtures.

Canada and the United States. With the rise of industrial powerhouses such as Brazil – part of the quartet of emerging superpowers along with Russia, India and China (the so-called BRICs) – and governments openly hostile to the United States (Venezuela, Ecuador, and Bolivia), the growing tendency is towards a hemispheric regionalism without the North.

This is not to say that North America does not have its regional allies – there are complex linkages weaving together material interests across states and there are many governments which share a common vision with the United States and Canada, even if the hemispheric pole of attraction may be increasingly shifting from North to South. One of the main arguments of this book is that the United States and Canada have increasingly strengthened their alliances with what I refer to as security-state polyarchies, which combine liberal-democratic institutions with heavy doses of repression, to maintain their commercial and geopolitical influence. This includes Mexico, Colombia, Haiti, and Peru. The focus on the agency of Canada and the United States in constructing a particular regional order that is increasingly being rejected, however, is meant to call attention to declining North American leadership in the hemisphere (though neoliberalism still very much remains entrenched).

For the United States, the political manipulation of democracy promotion in support of North American-led regional order is a continuation of longstanding forms of intervention; democracy promotion has long been used as a license to meddle in the domestic affairs of others. As authoritarian regimes across the Americas faced a crisis of legitimacy in the 1980s, many popular movements, governments, and parties were undermined by democracy promotion programmes in what we might refer to as the old democracy wars, which were intended to maintain traditional class structures during the period of democratic transition. The Sandinista government in Nicaragua was the most notable victim of this phase of interventionism. Democracy promotion was also used – unsuccessfully – to prevent the leftist liberation theologian, Jean-Bertrand Aristide, from winning the presidential election in Haiti in 1990, and to help build the democratic opposition against him during his years in exile following a coup in 1991. In Chile, democracy promotion helped build moderate political forces against more radical elements in the late 1980s to craft the transition from the Pinochet dictatorship (Robinson 1996). In the current phase of conflict, the focus is on countering left-wing 'populists' in countries where the transitions long ago took place, a political and ideological objective which is often explicitly formulated in the statements of policymakers, officials and programme documentation as we will see throughout this study. There is thus a high-level of cohesion between democracy promotion practices and other foreign policy objectives such as safeguarding free markets, fighting terrorism, and waging the War on Drugs.

Canada's political orientation has traditionally been less obvious, and many of the non-state actors that comprise its national field of practice in the area of democracy assistance have exercised a greater degree of autonomy from foreign policy pressures than their US counterparts. The analysis thus seeks to avoid reducing Canadian and US foreign policy actions to the same set of motivators and constraints – this book is as much about describing their differences as it is about revealing their similarities. Yet, Canada's approach has increasingly converged with that of its neighbour's to the south, and one of the main differences between the democracy wars against contemporary left governments and those of the past is Canada's involvement on the side of the United States. This has occurred as the United States has encountered a growing backlash against its democracy promotion activities that has placed new limits on its ability to use its programmes to accomplish regime change. Canada's ideological approach provides North American-led regional order with new hegemonic resources.

The arguments developed in this book draw upon and contribute to the ideas of critical researchers, particularly in the neo-Gramscian tradition of International Political Economy (IPE) and the work of William Robinson (1996, 2006), who argue that US democracy promotion contributes to the hegemony of the free market while stifling or co-opting visions that call for a greater degree of popular control over the economy. It addresses the main theoretical and empirical gaps of this literature, which has largely failed to consider how different Northern states advance different approaches to democracy assistance. For both the United States and Canada, the book distinguishes between democracy promotion strategies that mobilize and reinforce the hegemony of specific elite sectors against ideological threats and those that seek to consolidate polyarchy at a more structural level. In this sense, democracy promotion is interpreted as a modality of intervention which encompasses both strategies of stabilization and destabilization depending upon whether friends or enemies are in power. In the case of Canada, strategies of destabilization are less common. The approach to stabilization, moreover, is more systemic and less partisan than the United States', focusing more on building the capacity of state institutions to diffuse conflicts than strengthening the legitimacy of specific governments or supporting their ideological projects. Many of the multilateral efforts discussed in passing in this book are similar in nature.

Canada also has a more grassroots tradition of democracy assistance rooted in its traditional development field of practice, which includes many NGOs that seek to accompany popular organizations in their political struggles. It is a legacy of an increasingly remote era of enlightened self-interested development policy, however, and one that has all but disappeared. As I will argue, moreover, this tradition has always been embedded in unequal relations between North and South, in which Northern NGOs get to affect local political outcomes based in some cases on ethnocentric or partial understandings of the situation in which they operate. To the extent that this tradition has contributed to positive change, it is increasingly being replaced by a more compliant network of NGOs directly aligned with the foreign policy of the state and the interests of Canadian multinationals.

The perspective thus breaks with much of the mainstream literature on democracy promotion and democracy assistance, which tends to focus mainly on questions of strategy, assuming that democracy programmes are necessarily beneficial to the 'recipients' of the countries in which they are implemented. It seeks to avoid the conventional language of the field of international development, which frames foreign policy relationships in terms of 'donor' and 'recipient' states without paying attention to the fact that those who provide aid are often the ones who support the predations of the global economy. It questions, moreover, the legitimacy of the democracy promotion enterprise – its implicit claims to cultural superiority, mythology, double standards, and the foreign policy objectives to which it is often explicitly linked by its state practitioners. One can imagine, for instance, the popular outcry in Canada or the United States had Venezuelan President Hugo Chávez decided to launch a democracy promotion campaign in the wake of George W. Bush's highly contested electoral victory in 2000, or Prime Minister Stephen Harper's prorogation of the Canadian parliament in December 2009 to avoid discussion on the handling of Afghan detainees. The double standards have not been lost on Latin American leaders.

Haiti, Bolivia, and Peru have been selected as case studies. A critical examination of the programmes of US and Canadian state agencies, non-governmental organizations (NGOs) and quasi non-governmental organizations (QUANGOs) involved in democracy promotion in these countries provides insight into the range of strategies and tactics associated with democracy promotion and their underlying political objectives and impact. Of particular interest are the transnational linkages that these actors have cultivated with civil society organizations and political parties in the countries examined.

Each of these countries has been a major recipient of US and Canadian democracy assistance over the last decade. Each is also symbolic of larger trends occurring throughout the hemisphere over the last decade. As the leftist 'pink tide' swept the continent in the 2000s, the neoliberal government of President Gonzalo Sánchez de Lozada in Bolivia, a staunch regional ally of the United States, was replaced by a left-indigenous government when Evo Morales (head of a union of coca growers) won the presidential election in 2006. Bolivia has since aligned itself with those countries considered to be the most radical in the region, particularly Venezuela, Ecuador and Cuba. As such, Bolivia provides an important case study to examine how democracy promotion responds to shifting political fortunes. Furthermore, with a strong and growing Canadian mining presence, Bolivia allows us to explore the extent to which democracy promotion has been guided by or subordinated to Canada's own material interests. In Peru, the governments of both Alejandro Toledo (2001–2006) and Alan García (2006–2011) were important allies of both Canada and the United States, whose multinationals possess significant interests in the mining sector. Along with Mexico and Colombia, Peru (until recently) has served as a bulwark against the pink tide (and a key ally in the war on drugs); it therefore provides a case study of how democracy promotion operates on friendly terrain. Haiti, which has suffered a long history of US domination, has experienced a perpetual political crisis over the last decade. Since President Jean-Bertrand Aristide was ousted for a second time in 2004, a massive UN peacekeeping mission has remained to stabilize the situation, even though the country has never been at war. Canada's emergence as an imperial power in its own rights is most apparent in that unlucky country. A recurring threat to regional order and stability, Haiti's state of poverty and dependence render it fertile ground for the worst forms of democratic interventionism.

The centre-left governments of South America such as Brazil, Argentina and Uruguay are not considered in this study, where democracy promotion efforts are, in any case, minimal. Although Central America is not considered in any depth either, Chapter 2 provides some comments on recent democracy promotion efforts in Honduras, where the US and Canadian response to the coup in June 2009 shed more light on the contradictions of this form of practice.

The primary material of this study consists of non-attributable interviews with representatives of civil society organizations that have received democracy assistance, international organizations, political parties, academics, and Canadian and US embassy officials. Interviews were conducted over a three-month period beginning in January 2009 in Lima, La Paz, Cochabamba, Santa Cruz, and Port-au-Prince. Additional interviews in Canada with representatives of government agencies and NGOs, as well social movements, were carried out in Montreal and Ottawa. This research was supplemented by an extensive review of democracy assistance programme documents in three languages, statistical data on democracy assistance aid and primary documents on US and Canadian foreign policy.

Chapter 1 examines how Canada and the United States have sought to construct a regional order in the Americas based on a combination of neoliberalism and polyarchy. It will introduce the theoretical approach and will provide an historical overview of US and Canadian foreign policy in the Americas. This chapter provides the historical context for questioning the inherent beneficence of democracy promotion by examining the paradoxes and contradictions associated with the emergence of regional order at the end of the twentieth century. It also distinguishes between the different fields of practice in the area of democracy assistance that have evolved in the United States and Canada. Whereas the former has always been subordinated to larger foreign policy objectives, the latter has historically enjoyed a greater degree of independence from the state, though, as I will argue in this book, this is rapidly changing. The chapter ends with a neo-Gramscian typology of democracy assistance practices developed in greater empirical detail in the case studies.

Chapter 2 examines macro-level trends and developments in North American democracy promotion in the new millennium. It situates these in relation to the pink tide in Latin America, examining how the United States and Canada have deployed strategies of coercion and consent to re-impose a regional order that is increasingly unravelling. It analyses how both countries have responded to threats to the regional order by adopting the discourse of the post-Washington Consensus, and by using democracy assistance to undermine ideological threats. In the case of the United States, the Bush administration actively pursued a strategy of regime change in the first half of the decade, which led to an unprecedented backlash against democracy promotion. The Liberal government of Paul Martin also articulated new interventionist strategies such as the right to protect which were used to legitimize regime change in Haiti. With the regional backlash, however, new constraints were placed on the use of democracy promotion, and both the Obama administration and the Conservative government of Stephen Harper have focused primarily on using democracy assistance to support regional allies. The Harper government has also undertaken a concerted effort to reorganize the democracy assistance field of practice to advance its foreign policy objectives.

Chapters 3, 4 and 5 then explore North American democracy promotion in the case studies. In Chapter 3, which examines Haiti, I argue that both the United States and Canada used democracy promotion and assistance to undermine the left-wing government of President Jean-Bertrand Aristide and to legitimize the de facto government that was installed after his ouster in February 2004. Both countries intervened on behalf of the country's unpopular elite to prevent a rupture with the established order and to assist them in the ongoing task of imposing neoliberalism by force. The paradox is that while Aristide was denounced by many in foreign policy circles for having encouraged gang violence and for presiding over a narco-state, the violence and insecurity that followed was incomparably greater for the vast majority of Haitians. Since 2006, US and Canadian democracy promotion has taken a softer turn but continues to be riddled with the contradictions of imposing neoliberal polyarchy on a highly unstable and unequal social formation. This chapter also makes the case that Canada's grassroots tradition has always been susceptible to ethnocentrism and opportunism as Canadian NGOs can engage in local political struggles with a partial appreciation of the political situation and zero accountability to the lives of the people their actions affect.

In Chapter 4 on Peru, I argue that US democracy promotion has largely been geared towards stabilizing the fragile neoliberal polyarchy that emerged in the wake of the elite-led transition back to democracy in 2000 under presidents Alejandro Toledo and Alan García. The United States has been particularly concerned by the growth of 'anti-systemic' forces threatening the neoliberal model, a term designating both radical indigenous movements and the left-wing nationalist party, the *Partido Nacionalista Peruano* (Peruvian Nationalist Party – PNP). Although Ollanta Humala, the leader of the PNP, won the most recent presidential election in June 2011, his record thus far indicates that he is unlikely to break with the inclusive neoliberal model established by his predecessors. Canadian programmes have been less partisan, though they too have contributed to the hegemony of the neoliberal state by building its capacity to diffuse social conflicts. Peru also represents the progressive dimensions of the Canadian grassroots approach and its linkages with popular civil society, however, as some NGOs have provided funding to local organizations engaged in popular struggles.

Chapter 5 documents how US democracy promotion can switch from 'soft' to 'hard' tactics depending upon the shifting balance of power between contending social forces. In the early 2000s, while the neoliberal state faced a growing crisis of authority under President Gonzalo Sánchez de Lozada, US programmes dovetailed with the government's own efforts to build its legitimacy. As the indigenous movement organized under the leadership of Evo Morales and the *Movimento al Socialismo* (Movement towards Socialism – MAS) made its way to power, however, US programmes began organizing conservative forces against it. For the United States, the MAS represented a threat to both its ideological vision of regional order and its approach to controlling the drug trade through coca eradication. At the same time, the backlash against US democracy promotion placed constraints on its repertoire of interventionist tactics. Canadian democracy promotion followed a similar pattern in Bolivia as in Peru. In both countries, recent developments suggest that the grassroots approach is being cast aside by a new network of NGOs willing to align itself with the interests of Canadian mining companies.

The concluding chapter provides some comments on the future of North American-led regional order – and the role of democracy promotion in reproducing it – in light of the political changes that are sweeping the Americas. As Latin American states continue to challenge the legitimacy of democracy assistance programmes – and a new balance of power emerges between states – the United States and Canada may find it increasingly difficult to resort to interventionist strategies to advance foreign policy objectives.

Map 1.1 Map of the Americas

Source: The University of Texas at Austin, Perry-Castañeda Library Map Collection (802532, RO2283, 11-96). Available at: http://www.lib.utexas.edu/maps/americas/americas_pol96.jpg.

Chapter 1
Constructing Regional Order

At the seventh summit of the *Alianza Bolivariana para los Pueblos de Nuestra América* (Bolivarian Alliance for the Americas – ALBA) in Cumaná, Venezuela, in April 2009, leaders from several Latin American and Caribbean nations met to discuss plans to introduce a new regional currency, the Sucre, to replace the dollar as the main trading currency. In attendance at the summit were the leaders of the ALBA nations – a regional cooperation organization fostering trade and integration based on solidarity and mutual aid – as well representatives of states interested in joining the alliance.[1] The meeting was intended to set the tone for the upcoming Summit of the Americas of the OAS. After the president of Cuba, Raul Castro, denounced his country's ongoing exclusion from the hemispheric body at the ALBA summit, others took the opportunity to send a message to Washington about the independence of the region.

Taking aim at the long-time policy of the United States of opposing Cuba's ascension to the OAS, Fernando Lugo, the new president of Paraguay, wondered who at the UN or the OAS had given the United States the power to elect itself as the judge of the democratic people of Latin America. 'We are in a new moment in Latin America', Lugo declared, 'and we are the true authors of our own destiny, and there is no nation in the world that can judge us, over who is more democratic or not. We each have our own unique process of democratization.' If Cuba was the focal point of the discussion, President Lugo's rebuke spoke to the region's new assertiveness. Bolivia's Evo Morales added further that: 'the United States doesn't have any authority to speak about democracy, because from over there they install coup d'états, like these civil coups now in Bolivia' (Fox 2009). Those in attendance understood that the Bolivian president was rehearsing a longstanding – and, we shall see, not unfounded – complaint that the United States was using its democracy promotion programmes to support politically-conservative forces in the departments of Bolivia's eastern lowlands, which had recently launched autonomy referenda to reclaim political power from the indigenous-led central government in La Paz.

The statements made by the leaders during the summit symbolized the new-left Latin American zeitgeist, a far cry from the days of the Washington Consensus when the promotion of free markets under US leadership was the order of the day. Many had risen to power in the wake of the regional backlash against neoliberalism in which social movements across the continent had mobilized in demand of a

---

1   The member states of the ALBA are Antigua and Barbuda, Bolivia, Cuba, Dominica, Ecuador, Nicaragua, Saint Vincent and the Grenadines and Venezuela.

more socially just and inclusive order. Although the rhetoric of some new left leaders has far surpassed their actual commitment to getting rid of neoliberalism, some have expanded popular control over key resources and have introduced new spaces for democratic participation that transcend the traditional liberal-republican order (see Cameron and Hershberg 2010). But the larger regional transformation is not just about ideology – with the establishment of the *Comunidad de Estados Latinoamericanos y Caribeños* (Community of Latin American and Caribbean States – CELAC) in February 2010, even US and Canadian allies like Mexico and Colombia recognize the advantage of deepening alliances with left and centre-left governments who are rejecting traditional forms of political interference. Though many governments continue to benefit from the old patterns of intervention – which bolster their own efforts at maintaining internal order – even the most conservative are weary of being left behind as the region undergoes a new wave of integration that excludes Canada and the United States (indeed, CELAC is essentially the OAS without these two countries). As the first decade of the new millennium drew to a close, such developments had transformed the political landscape, shifting the hemispheric balance of power away from Washington, consolidating new regional visions of economics and politics, and promising a future of greater regional pluralism. The unanimous decision to readmit Cuba into the OAS at the Summit of the Americas in Trinidad that followed on the heels of the ALBA meeting symbolized the waning ability of the United States to dominate the region politically.[2]

Two and half months after the Summit, however, when the president of Honduras, Manuel Zelaya, was forced to flee his country by the military on the spurious charge that he had violated the constitution, the new sensibility seemed to be in jeopardy. Although every country in the region demanded the immediate return of Zelaya, the United States and Canada acted with reticence, condemning the coup only to legitimize the show-election held by the de facto government in conditions of extreme repression as most of Latin America stood aghast. Both the government of Stephen Harper and the administration of Barrack Obama had prioritized democracy in their regional foreign policy, and many wondered whether commercial interests were trumping idealism as the new conservative government in Honduras re-opened for business. The events suggested that the Latin American right had not forsaken the authoritarian tactics of the past, and that the two countries which had long claimed the mantle of democracy in the hemisphere were once more not as idealistic as they purported to be.

The struggle over the meaning of democracy – and who possesses the legitimacy to promote it – constitutes the political backdrop of the investigation in the pages that follow. There is nothing new about the incongruence between stated objectives in support of democracy and actual foreign policy practice, however,

---

2   To date, however, Cuba has not applied for readmission. Consequently, it was not invited to attend the upcoming 2012 Summit of the Americas in Cartagena, Colombia, by its host, President Juan Manuel Santos.

and there is good historical reason to insist on the necessity of a critical analysis of US and Canadian democracy promotion in the Americas. Those who know the history of US-Latin American relations are well aware that past authoritarian regimes throughout the region were nearly always allies, and that the historical record is replete with examples of successive US administrations undermining left-wing movements and governments that interfered or threatened to interfere with US interests. US democracy promotion has often been contradictory and self-serving, largely subordinating any genuine concern with supporting the will of the people to the imperative of maintaining free markets. Washington is not the only one, however, as Ottawa increasingly uses democracy promotion as a rationale for advancing security and commercial objectives that are intertwined with those of the region's traditional hegemon. To be sure, this does not signify a radical break with the past: the Washington consensus on the desirability and necessity of neoliberal policies was ardently supported by Canada throughout most of the 1990s as its multinationals, particularly in the extractive sector, greatly expanded their operations. Yet in the construction of North American-led regional order, Canadian democracy promotion has never been as explicitly political as its neighbour's. While the Canadian state has tied its support to democratic norms in the interstate system to its economic agenda, the field of practice that provides democracy assistance has never been as directly and uniformly subordinated to the undemocratic foreign policy objectives of state. In fact, there are many actors in the Canadian democracy assistance field of practice that do genuinely support bottom-up democratic development. We will see in the following chapter, however, that this is rapidly changing.

The chapter is divided into three sections. The first section provides a brief review of dominant themes in the literature on democracy promotion, arguing that mainstream accounts insufficiently problematize the tension between the concurrent promotion of democracy and neoliberalism by the United States and Canada. I then develop an alternative historical approach to understanding US and Canadian democracy promotion in the Americas based on the concept of regional order. Here, the focus is on the shared affinity of the United States and Canada for regional forms of governance rooted in neoliberal polyarchy. Section two then examines how US and Canadian approaches to democracy promotion developed within very distinct fields of practice despite this shared affinity; the emphasis is thus on the key differences that mark the approaches of different actors who deliver democracy assistance. The final section provides a neo-Gramscian theorization of how different democracy promotion strategies produce hegemony and, in some cases, counter-hegemony. This provides the historical and theoretical background to the investigation in the following chapter, which focuses on how US and Canadian democracy promotion has evolved and responded to threats to neoliberal regional order in the new millennium as the immediate backdrop for the analysis of North American democracy promotion in Haiti, Bolivia and Peru.

The Authoritarian Origins of Neoliberal Polyarchy

The Limitations of Mainstream Accounts

Throughout most of the twentieth century, the United States sought stabilization in countries threatened by popular revolt and destabilization where popular movements or leaders came to power. These modalities of intervention have long been two sides of the same coin.[3] This is the background to US foreign policy in the region, and it is within this historical legacy that we must situate the emergence of democracy promotion as a foreign policy practice in the hemisphere. Although Canada's engagement with the Americas has been quite different, the construction of a regional order predicated on neoliberal polyarchy was rooted in the imperial dynamics of the past whose legacy continues to shape North-South hemispheric relations. Situating the analysis that follows within this legacy identifies the continuities that inhere in democracy promotion as a foreign policy practice, and provides a counter-narrative to theoretical and policy approaches to democracy promotion that gloss over, justify or conveniently forget this past.[4]

US narratives of democracy promotion have traditionally revolved around whether the United States should actively promote democracy or simply lead by example according to its own perceived unique national destiny – what Secretary of State Henry Kissinger referred to as the crusader versus the beacon approach to foreign policy. Both 'vindicationalist' (the idea that the United States should act as an agent of democratization) and 'exemplarist' (the view that it focus on perfecting its own democracy as an example to others) arguments, however, have constructed US national political identity around the concept of 'exceptionalism,' which has historically held that the United States differs qualitatively from other developed nations because of its unique liberal-democratic credo (a concept steeped deeply in the Puritan past). If the exemplarist perspective held sway throughout most of the nineteenth century – and provided a powerful frame that guided the foreign policy views of the founding fathers of the republic – the emergence of the United States as a great power in the 1890s provided policymakers with the material capabilities to pursue a more interventionist foreign policy. The debate would continue

---

3   They are well documented in Eduardo Galeano's historical classic, the *Open Veins of Latin America*, a copy of which Venezuelan president, Hugo Chávez, presented to Barack Obama at the OAS summit in Trinidad.

4   Those writing about democracy promotion are faced with the difficult task of having to sift through an incomparably vast literature that straddles multiple social sciences, crosses disciplinary divides and includes conversations between scholars and policymakers and among scholars themselves. Here, I approach the subject from the 'top-down' vantage point of international relations to comparatively assess what has been said about US and Canadian approaches before building an alternative critical framework rooted in international political economy. In developing this framework, I draw upon contributions in comparative politics and historical sociology that focus on both the conditions conducive to democracy and the actors involved in the democratization process.

throughout much of the twentieth century, with both Democratic and Republican administrations alternating between more activist and isolationist foreign policies, though the post-war era would usher in a generalized commitment to building a liberal-democratic order. Despite their aversion to the constraints imposed by liberal multilateralism, even realists argued that US nationalism had traditionally been defined in terms of both the country's adherence to a set of liberal, universal political ideals and a perceived obligation to spread those norms internationally in the interests of both the United States and the rest of the world.[5] Breaches in the international democratic ethos were justified in terms of the exigencies imposed by the global struggle against communism.

During the Reagan administration, a particularly aggressive variant of vindicationalism, neo-conservatism, reoriented US foreign policy around a democratic crusade against communism (discussed later in this chapter) which left some more traditional liberal internationalist deeply uncomfortable. With the end of the Cold War, however, democracy promotion advocates of all stripes argued that the demise of the Soviet Union opened up the possibility for the consolidation of a liberal-international order rooted in the spread of free markets and liberal-democratic institutions. Neoconservatives like Joshua Muravchik (1992) argued that democracy was the quintessential American ideal, the spreading of which was America's destiny. With the 'third wave' of democratization that swept Latin America and the former Soviet-bloc countries, liberal scholars and democracy practitioners applauded the 'democratic enlargement' policy of the Clinton administration as a welcome break with the past, when autocratic regimes were tolerated (if not actively supported). In the Americas, proponents argued, the US shift in foreign policy was a key factor in contributing to the development of an interstate regime founded on the collective defence of democracy (Parish and Peceny 2002). The benefits of supporting authoritarian governments had changed, many argued, and the spread of democracy would help secure international peace (a claim based on the theory and empirical observation that two democracies had never gone to war (M. Cox 2000)).

Throughout the 1990s, practitioners such as Thomas Carothers advanced powerful arguments on the importance of promoting democracy abroad while warning that the results could be slow, and that 'democracy promotion must be approached as a long-term, uncertain venture' (Carothers 2004: 351). Others warned that the overriding imperative should be on consolidating those democracies that came into being during the third wave (Diamond 1999), and cautioned against accepting the results of flawed elections in countries where democracy remained weakly consolidated (Schraeder 2003).[6] Although much of the literature struck the

---

5 See Monten (2005) for an overview of US exceptionalism and the mainstream debates that have historically characterized the discussion on US democracy promotion.

6 With most scholars sharing a normative preference on the desirability of promoting democracy, much of the literature turned to the efficacy of specific democracy promotion strategies (see Ethier 2003, for example, for a comparative analysis of democracy promotion

chord of traditional Wilsonian idealism, liberal advocates held basic assumptions on the goodness of US intentions in spreading democracy abroad in common with more ideologically-pure neo-conservatives.

With the advent of the Bush Doctrine (discussed in the next chapter), neo-conservatives took on a more prominent role in both academic and policy circles, arguing that the right to intervene was embedded in the superiority of American democracy and its privileged international power. In this formulation, the alleged relationship between democracy and security also figured prominently. Interventionism was to be applauded because it is exercised on behalf of democratic and liberal ideas (Karatnycky 2004) or towards humanitarian ends (an argument also made by Harvard-based Canadian intellectual and former head of the Liberal Party, Michael Ignatieff (2003)). If relative power shapes the basic parameters of a state's foreign policy, as realists contend, unipolarity had created a more permissive environment in which an aggressive ideology of democracy promotion could flourish (Monten 2005). Many liberal internationalists, however, were critical of the new interventionism, and called for a more careful and internationally-coordinated approach (Carothers 2006, 2009).

In Canada, the democratic transitions of the 1980s and the collapse of communism opened up discussions that resonated with the liberal internationalism of many US scholars albeit with a distinct Canadian hue. Although Canada's role in the post-war order was often idealized from the perspective of a Canadian exceptionalism that downplayed the state's economic and security interests and emphasized Canada's contribution to multilateralism, some scholars took note of the strategic value of fostering a more cooperative, liberal international order (Cooper, Higgott, and Nossal 1993). Policymakers considering an enhanced role for Canadian democracy assistance and support for human rights were weary of the neo-conservative approach, and sought to define a Canadian perspective based on this more collaborative conception of global cooperation (Schmitz 2004). Although the promotion of democracy in itself did not figure prominently in foreign policy statements of the 1990s, Canada championed a human security agenda that grouped together a broad range of rights.

During the Liberal governments of the early 2000s, discussions centred on the need to expand democracy assistance efforts and more consistently uphold democratic values on the international stage (notably Axworthy, Campbell and Donovan 2005). In one influential series of contributions, Sundstrom (2005) noted that democracy had traditionally been framed as a 'Canadian value' and that there was intrinsic value to supporting it abroad, while Perlin (2003–04) invoked the theory of the democratic peace to build a more instrumental case in support of an enhanced Canadian role. International relations theorist Jennifer Welsh (2007, 2003) argued that Canadians were uncomfortable about imposing values of democracy,

---

efforts, including Canada's), with even the more sceptical conceding that, if nothing else, democracy assistance as a whole seemed to have some impact on mitigating authoritarian backsliding (an argument made by Scott and Steele (2005) in the context of Chávez's Venezuela).

rule of law and human rights on others, but that these values were in fact shared by several members of international society. Andrew Cohen's (2003) popular book, *While Canada Slept*, argued that Canadian foreign policy was too modest and self-effacing, and that Canada should offer itself as the good governance nation (an argument that combined a uniquely Canadian exemplarist and vindicationalist approach). Appearing before parliament, Michael Ignatieff called for 'peace, order and good government' as an organizing frame for Canadian activities because it encapsulated the unique tradition Canada had to offer: democratic institutions, federalism, minority rights guarantees, linguistic pluralism, aboriginal self-government and a positive enabling role for government in economic and social development (Schmitz 2004). As Conservative governments began to chart a new unilateralist foreign policy (discussed in the following chapter), Canadian politician and Liberal Member of Parliament, Bob Rae (2010), restated many of the same arguments in his book, *Exporting Democracy*.

Like the United States, Canada's role in support of inter-American democratic institutions – where its democracy promotion efforts have been most concentrated – has received particular positive attention (for example, Major 2007). Indeed, Canadian and US efforts in supporting democracy both in the Americas and the rest of the world have been seen to be complementary. Together, according to Scott and Walters (2000), Canadian and US actors had formed a 'transnational democracy issue network' promoting values which were becoming international norms in a new world order built on liberal democratic government and global civil society.

Despite considerable differences within the literature, most mainstream accounts of US and Canadian democracy promotion implicitly or explicitly assume that free markets and democracy are mutually reinforcing (though the Canadian literature is generally more social democratic than the American) and that both countries are genuinely contributing to a more liberal and humane international order (though foreign policy strategies and tactics are often in dispute). Many explicitly conflate the spread of democracy with the spread of capitalism,[7] while others largely remain unconcerned by the explosion of inequality that has accompanied the democratic transitions and the deepening of market relations (or at least do not call attention to the contradiction of foreign policies that support both processes). For Canada, the

---

7   The relationship between capitalism and democracy should be situated historically based on recognition that the principles of equality embodied in the concept of democracy and the drive toward individual accumulation have always been in conflict. As Barrington Moore (1966) argued long ago, the liberal bourgeois transformations of Western Europe were neither democratic nor egalitarian except in the narrow sense that they called for constitutional governance by a new commercial class against aristocratic and monarchic privilege (see also Wood 1995). The extension of the suffrage in western capitalist states only occurred gradually and unevenly as social struggles forced governing elites to make concessions to working class and suffragette movements. Capitalism did not give rise to political democracy in simple gradualist fashion but only after traversing a long historical road fraught with violence and crisis in which market forces were progressively constrained (Therborn 1977).

focus on its distinct cooperative approach does highlight an important difference that has shaped its traditional democracy assistance field of practice (as we shall see), but it equally obfuscates important foreign policy similarities. Abstracted from a larger critique of the role of Canada and the United States in the international political economy, advocates and practitioners alike overlook how the neoliberal model promoted by both states is inimical to democratic development (discussed further below). Needless to say, critical investigations such as Robinson's (1996) that bring to light how democracy assistance itself can be used to support dominant classes and social groups that perpetuate highly unequal social orders have received scant attention within more mainstream circles. This is not to suggest that there is no critical literature on US and Canadian democracy promotion. As we will see in the following pages, there have been some important contributions, though a comparative regional account is conspicuously lacking.

Narratives of regionalism in the Americas also tend to overlook how democracy promotion can be used to reinforce class relations across states in profoundly undemocratic ways. For most, the evolution of regionalism is the story of waning US hegemonic domination in terms of its asymmetrical power and the rise of a more multipolar regional order (see, for example, Mace et al. 2011). Those focusing on democracy have examined the inter-American democracy promotion regime institutionalized by the OAS (Legler 2011) rather than the North-South hemispheric relations that determine who promotes democracy and in whose interest. Few have examined the history of democracy in the Americas as it has evolved in relation to North American material interests, regional class alliances, and contending political-ideological projects. Viewing regionalism from this perspective calls attention to the nature of neoliberal polyarchy, its contradictions and its structural continuities with authoritarianism. To approach this task, the remainder of this chapter will develop a critical theoretical framework for exploring North American democracy promotion through an historical analysis of the continuities and contradictions of foreign policy in the Americas and the evolution of democracy assistance as a field of practice. The methodological approach will combine history and theory as the concepts used to interpret the behaviour of state and non-state actors are developed within the historical narrative of regional order.

The History of Regional Order as a Critical Alternative

Critical IPE – or historical materialism – provides an alternative framework for conceptualizing the internal and external dynamics of democratization that focuses on social structures, transnational relations, and class and state power. Scholars in the tradition of critical IPE have long followed Robert Cox in examining the intersection of structural, institutional and ideological factors in the making of world order, understood as the relations of wealth and power in the world economy and international system (Cox and Sinclair 1996). In any world order, a dominant mode of production penetrates and subordinates economies such that extraction of surplus flows from weaker levels of production to stronger ones. Although the point of departure in this analysis is the world capitalist system (or, more recently, global capitalism), the regime of accumulation supported by the United States in the Americas was intricately tied to this system while still possessing its own distinct historical geography rooted in the transnational class relations of the region and the preponderant role of the United States.

The concept of regional order has been pursued by the critical theorist Anthony Leysens (2008), who follows Gamble and Payne (1996) in arguing that regionalist projects in the last few decades have largely articulated with the global economy under US leadership through a form of 'open regionalism'. While the phases of regionalism in the Americas will be discussed below, it bears mention here that Leysens emphasizes the importance of looking at the material, ideological and institutional dimensions of regionalism, and in exploring both the role of states and social forces in supporting regional projects. This will be accomplished in the following pages through a critical look at the ideological dimensions of North American democracy promotion and the linkages fostered between, states, state agencies, and civil society through democracy assistance programmes. Regionalism in this sense is viewed as a historical process led by state and non-state actors that brings into being historical structures rooted in new transnational class alliances and evolving state forms.

The evolution of regional order in the Americas can be viewed in terms of three distinct phases: 1) the Pan-American system launched at the First International Conference of American States in 1889 which lasted until the end of the second world war, during which time the United States displaced Britain as the main imperial force in the region; 2) the emergence of the modern inter-American system institutionalized through the OAS during the post-war period and; and 3) the most recent phase of open regionalism driven by hemispheric economic integration ushered in by the first Summit of the Americas in Miami in December 1994 (Mace, Cooper and Shaw 2011). What follows is a brief account of these three phases, with a particular focus on the latter two.

Although the US ambition to establish a sphere of influence in the hemisphere free of European interference was declared early in the nineteenth century through the Monroe Doctrine (1823), the International Conference of American States in 1889 marked the first concerted effort to articulate a Pan-American regional order under US leadership. Among other things, the Conference, which was held in Washington and attended by 13 Latin American nations, led to a series of commercial and trade agreements and the creation of permanent secretariat that would later become the Pan American Union (and ultimately the OAS). The new US regional leadership coincided with the expansion of US interventionism in the region, justified from the outset in terms of securing the conditions for freedom and democracy to flourish (for example, the US intervention in the Cuban War of Independence from Spain, which led to the Spanish–American War in 1898 (see Monten 2005)). Yet, beyond the rhetoric, such escapades were early instances of a new pattern of interventionism intended to establish mutually advantageous arrangements between US capital and local dominant classes.

Interventions in support of authoritarian state structures seeking to maintain order against popular revolts became a regular practice in the early twentieth century, particularly in those regions over which the United States exercised the most control and where its economic interests were most entrenched. US material interests included maintaining access to raw materials (oil, iron, bauxite, copper, and tin), agrarian products (sugar and coffee), markets for US products and safeguarding important trade routes. Between 1904 and 1934, the United States undertook more than thirty interventions and occupations in Central America and the Caribbean (see Leogrande 1998), including a nineteen year occupation in Haiti (1915–1934) (see Castor and Garafola 1974). US capital also expanded throughout the mining industry of South America and brought with it new forms of political interference in that sub-region as well. In the Andes, US capital targeted two growing sectors in the mining industry – copper in Peru and tin in Bolivia.[8]

Canada's role in the regional order of the period was minimal. Its industrial base and integration into a continental labour market with the United States prevented the more aggressive forms of US domination that developed throughout the rest of the hemisphere. As an advanced capitalist democracy in its own right, its subordination to US imperialism was sustained through cultural hegemony in civil society and class struggles channelled through liberal democratic institutions (Williams 1989). As for its relationship with the rest of the region, Canadian commercial interests in Latin America were modest, though Canadian banks did predominate in the English Caribbean throughout much of the late nineteenth and most of the twentieth century. The Canadian state protected such interests vigorously, and in many cases sent troops to help British authorities re-establish order during periods of political unrest (examples include Bermuda from 1914 to 1916, St. Lucia from 1915 to 1919, and Jamaica from 1940 to 1946) (Engler 2009).[9] Canada also established a small presence in the mining sector in the Andes when Canadian-based capital took over the British company, Peruvian Corporation (Becker 1983). Overall, however, Canada largely resisted the hemispheric pole of attraction.

In the post-war period, the beginnings of a new regional order were inaugurated at the Ninth International Conference of American States, held in Bogota in 1948, which led to the establishment of the OAS. The new institutional architecture included a defensive military alliance, the Rio Treaty, the Inter-American Development Bank (IDB), and various democracy and human rights institutions including the Inter-American Convention of Human Rights (IACHR), the Inter-

---

8 When economic recession hit in the mid-1920s, US banks began overseeing the collection of Bolivia's taxes and custom receipts and played a direct role in setting government fiscal policies (Lehman 1999). In Peru, the US multinational, Cerro de Pasco Copper Corporation, came to dominate the mining industry for the first half of the twentieth century. By the 1930s, the company became Peru's single largest landowner (Becker 1983).

9 Interestingly, the Canadian military had plans to intervene in Jamaica in 1979 to protect the alumina-processing facilities of the Montreal-based company, ALCAN, from political unrest. The plan never came to fruition, however (Maloney 2000).

American Commission on Human Rights and the Inter-American Court of Human Rights. Despite the new democratic norms enshrined in these institutions, US power was still projected in support of authoritarian state forms geared towards the mutual interests of Latin American oligarchs and US capital. US allies tended to preside over social orders that governed through repression – in the language of Antonio Gramsci, they were non-hegemonic.

A brief theoretical digression is in order to explain the significance of this concept, which is used throughout this book to explore the construction of power relations both within and across states. Within a particular social formation, hegemonic representations of society are constructed socially by organic intellectuals associated with different social groups and classes and confer considerable legitimacy upon the established order. The hegemony of a dominant class produce 'imagined communities' that stabilize the structure of accumulation and rationalize social contradictions by obfuscating structural antagonisms or transforming them into simple differences (Persaud 2001). Hegemonic representations are based on 'common sense' understandings of the world that speak to people's everyday experience at a superficial level of existence. They become socially embedded to such an extent that they are taken to be the natural order of things (Cox 2001). They are articulated and defended across the multifarious institutions of civil and political society (or the integral state) within realms such as the educational system, state apparatuses, the church and the public media through a totality of institutional and discursive practices that constitute who we are and how we think (Williams 2005). Applying the concept of hegemony to the international sphere, neo-Gramscians have argued that hegemonic world orders are underpinned by historic blocs, or alliances of social classes, across states whose power and authority are considered legitimate. Their rule is cemented through universalistic ideas, as well as a certain degree of material concession to subordinated classes. *Pax Americana* in terms of the relationships between core states of the world capitalist system in the postwar period was indeed hegemonic; in terms of the relationship between the United States and the periphery, however, force prevailed.

Thus, when traditional patterns of oligarchic rule were threatened – in the wake of the presidential victory of the leftist Jacobo Árbenz in Guatemala in 1951, for example, or the Bolivian Revolution of 1952, or the epic victory of Salvador Allende in Chile in 1971 – the United States reacted swiftly to restore the power of dominant classes through support to conservative military factions (Blum 2004). CIA covert operations largely replaced the direct interventions of the past as the new form of interventionism (though interventions still occurred, such as when the Marines landed on the soil of the Dominican Republic to quell unrest in April 1965, marking the beginning of an occupation that lasted over a year). Institutions like the infamous School of the Americas at Fort Benning, Georgia, provided training to Latin American security forces in counter-insurgency tactics, including some of the worst human rights offenders in the region.

In much of South America, the working class did acquire growing political power in the post-war period exemplified by the national-development states of

Juan Perón in Argentina and the military government of Juan Velasco in Peru (the reach of the US imperium was less powerful in the more advanced economies of the *Cono Sur*). Reformist initiatives like John F. Kennedy's Alliance for Progress also sought to soften the edges of imperial power by promoting weak land reform initiatives, though these coincided with the extension of covert operations. An initial wave of integration through preferential trade agreements such as the *Mercado Común Centroamericano* (Central American Common Market), the *Asociación Latinoamericana de Libre Comercio* (Latin American Free Trade Association), the Andean Pact and the Caribbean Community and Common Market (CARICOM) also provided for a measure of South-South cooperation focusing on market access in goods (Estevadeordal and Suominen 2011).

During the Nixon administration, the United States played a key role in supporting the Pinochet coup in Chile which ushered in the Dirty Wars that restored class power through the military dictatorships of the 1970s. Through Operation Condor, the military governments of Argentina, Chile, Uruguay, Paraguay, Bolivia and Brazil, formed a regional alliance to carry out combined extraterritorial operations using disappearance, torture, and extrajudicial execution to eliminate political enemies with US support (McSherry 2005). Such policies underpinned a US-Latin American interstate regime founded upon calculated and systematized political violence largely intended to maintain social inequalities (Rodriguez and Menjívar 2005). In Central America, where guerrilla forces emerged to challenge repressive authoritarian regimes in Guatemala, El Salvador, and Nicaragua, the Reagan administration (1981–1989) launched devastating counter-insurgency wars supplemented by the new modalities of democracy promotion (discussed below). This approach reversed the brief opening that had occurred under the Carter administration (1977–1981), which reduced aid to countries violating human rights gave new impetus to the burgeoning democratization movements across the hemisphere (Pastor 2001).

As for Canada, it continued to avoid regional politics and US-dominated institutions such as the OAS throughout the post-war period, largely respecting the US sphere of influence (Klepak 2006). Canadian governments, particularly the Liberal governments of Pierre Trudeau (1968–1979, 1980–1984), maintained a certain degree of foreign policy autonomy, however, breaking ranks with the United States on key foreign policy issues in the region such as Canada's position towards socialist governments like Cuba and Nicaragua under the Sandinista government of Daniel Ortega, with which it maintained positive relations (Matthews and Pratt 1988). Although Canada enjoyed a privileged position in the world capitalist system, the formulation of foreign policy was arguably open to a broader stream of currents, including a more radical approach to international development assistance rooted in what Pratt (2003) refers to as humane internationalism, which included the conviction that an orderly and more equitable world was in the interests of the rich industrialized countries.

The (Partial) Democratic Transition

Three important shifts began to occur in the late 1970s that transformed the regional order: 1) nearly all Latin American states underwent significant neoliberal restructuring; 2) polyarchy began to gradually replace authoritarianism; and 3) by the 1990s, Canada emerged as a regional force in favour of the new order. The first two factors were closely interrelated. Both were linked to the debt crisis, which overlapped and interacted with larger structural trends in the world economy. This included declining profitability as the multinationals of advanced capitalist states became increasingly competitive, increased oil prices (the oil shocks of 1973 and 1979), combined inflation and unemployment (stagflation), and the demise of the Bretton Woods system. These structural transformations led to a new regime of accumulation based on neoliberal globalization as a method of restoring profitability (Harvey 2010).

Fuelled by the infusion of petrol dollars into the international banking system, private banks and international financial institutions began to lend profligately to Third World Countries in exchange for economic reforms. The pioneers of the reforms were the military governments of South America (particularly Chile) and Haiti under the dictatorship of Jean-Claude 'Baby Doc' Duvalier. But the debt crisis and the economic turmoil of the late 1970s and early 1980s also contributed to a crisis of authority for the authoritarian regimes, including those which had spearheaded the first cycle of neoliberal reforms. As collective actors re-emerged to demand democratic change (Rueschemeyer et al. 1992, Neuhouser 1998, and Collier and Mahoney 1997), liberal elites began supporting polyarchy as an alternative form of governance to both authoritarianism and deeper notions of democracy, one with greater potential to manage social conflicts. Elites and militaries formed pacts guiding democratic transitions that promised to leave the economy and its attendant class relations untouched while providing amnesties for the perpetrators of terror. While the adoption of polyarchy was an historical development representing considerable progress from the days of the military dictatorships, it also reflected these profound limitations.

Such elites received support from core capitalist states and transnational institutions which began promoting low-intensity democracy as the political flipside to the emerging global economy (Robinson 1996). Latin America's new leaders were reinforced by a never ending cycle of borrowing from the international financial institutions, conditional upon adjustment programmes which reduced the public sector and further integrated the national economy into global capitalism. This coincided with more comprehensive regional and sub-regional trade and investment agreements – some of which replaced earlier arrangements – such as the *Mercado Común del Sur* (Common Southern Market – Mercosur) and the *Comunidad Andina* (Andean Community, which replaced the Andean Pact) (Estevadeordal and Suominen 2011). Deregulation of the financial sector and the removal of capital controls, liberalized trade agreements and investment regimes, among other policies, led to the subordination of the national economy to transnational capital, particularly its speculative financial component. In terms of production, neoliberalism prioritized primary exports for external markets in the name of comparative advantage and to accumulate foreign exchange reserves. This policy orientation enhanced the class power of new and old sectors of the agro-oligarchy and bankers and merchants integrated into the global economy (Petras and Veltmeyer 2009).

The combination of neoliberalism and polyarchy rested on two fundamental paradoxes that point to the contradictions of the emerging order. First, just as the democratic transitions were occurring, neoliberalism was largely being achieved through undemocratic means. As political elites began contesting elections for the first time in years, victorious leaders began presiding over neoliberal stabilization packages in direct violation of their campaign promises. Such was the case in Argentina under the presidency of Carlos Menem (1989–1999), and in Venezuela under Carlos Andrés Pérez (1989–1993) – both of whom were elected on social democratic platforms (Silva 2009).

In Bolivia, elite political parties formed power-sharing pacts in the wake of the democratic transition in 1982 premised, in part, on a shared commitment to neoliberal reform. In 1985, the Bolivian government under President Gonzalo Sánchez de Lozada transformed the economy through a single decree with 220 separate laws. In Ecuador, President Febres Cordero (1984–1988) implemented neoliberalism by decree in the face of congressional opposition. This trend persisted throughout the 1990s in several countries, including Peru, where Alberto Fujimori decreed neoliberal reforms after winning an election in which he campaigned against economic adjustment. In Haiti and Nicaragua, polyarchy only took root once more participatory democratic experiments were quashed through a combination of military and para-military repression, complemented by US democracy promotion programmes that bolstered elite social forces.

The United States also began to articulate new rationales to justify counter-insurgency tactics to stabilize capitalist allies as the Cold War came to a close. In 1989, the Bush administration launched the Andean Initiative, a five-year plan that made Bolivia, Peru and Colombia the leading recipients of US military aid in Latin America in a vast expansion of Reagan's regional War on Drugs. As Stokes (2006) documents in his important book on America's war in Colombia, military aid was conditional upon neoliberal economic reforms. The US Congress passed Section 1004 of the National Defense Authorization Act in 1991, reorienting US military aid and training towards the new war on drugs, with the corollary benefit of helping democratic governments fight growing leftist insurgency. Although US defence agencies were apparently aware that targeting guerrilla insurgents would do little to disrupt the drug trade, Colombia became the third largest recipient of US military aid, with the military and paramilitaries (themselves often linked to the drug trade) concentrating their efforts on eliminating the guerrilla *Fuerzas Armadas Revolucionarias de Colombia* (Armed Revolutionary Forces of Colombia – FARC).

In short, with the economy remaining safely in the hands of powerful business interests and protectionist barriers preventing the further penetration of transnational capital undemocratically removed, most countries underwent a process of regime change rather than structural transformation. This in large part accounts for the social struggles that continue to define Latin American politics as new left governments have come up against the constraints imposed by an economically-powerful resurgent right in countries such as Bolivia, Venezuela, Ecuador, and Honduras (Petras and Veltmeyer 2009) (discussed in the following chapter).

The second major paradox was that neoliberal reforms undermined the very social basis of democracy. As much of the literature on democratization has demonstrated, particularly the work of Muller (1995), high levels of inequality can prevent or undermine democratic consolidation. In Latin America – historically one of the most unequal regions of the world – indicators of income concentration remained unchanged or worsened between 1990 and 2002 (ECLAC 2004). Although neoliberalism dealt with the hyperinflation of the 1980s, it performed poorly on most other indicators. With massive layoffs in the public sector, wages stagnated throughout the 1990s. Between 1990 and 2008, annual per capita GDP growth was a meagre 1.7% in the region, well below the rate recorded in East Asia (4.1%). GDP grew considerably less than in the 1970s, and only experienced a significant annual growth rate of 5.3% during the five-year period from 2004 to 2008 (ECLAC 2010a). The increase during this period, in turn, reflected an improvement in terms of trade rather than a complete break with the neoliberal model (Petras and Veltmeyer 2009).

The region also underwent a profound transnationalization of its production and service infrastructure associated with the wave of privatizations and removal of barriers to speculative finance capital. As Robinson (2004) points out, Latin America was a net exporter of $219 billion in capital surplus to the world economy during the 'lost decade' of 1982 to 1990, and then became a net importer from 1991 through to 1998. During this period, nearly $830 billion in capital was transferred to the region primarily in diverse portfolio and financial ventures, such as new loans, the purchase of stock in privatized companies, and speculative investment in financial services, such as equities, mutual funds, pensions, and insurance. After this initial influx, the region once again reverted to an exporter of capital starting in 1999. Latin America also began a process or relative de-industrialization; the share of manufacturing as a percentage of GDP declined from 27% in 1980 to 17.9% in 2009 (compared to 31.4% for East Asia and the Pacific) (Paus 2011). Commenting on the three patterns of linkage with the global economy – one based on natural resources for South America, another based on *maquila* activities for Mexico and Central America, and the other based on services for the Caribbean – ECLAC (2009) warned that 'the degree of articulation with the local productive apparatus has been unsatisfactory, at the detriment of the development of national suppliers and endogenous technology capabilities. On the contrary, the "opening-up" process, together with higher import contents, has tended to reduce linkages that existed prior to trade liberalization.'

In addition to the international financial institutions, the economic and political aspects of the neoliberal project were institutionalized in the regional system by the OAS. Just as the IFIs began advancing a notion of 'good governance' in the 1980s, the OAS began championing liberal democratic norms as a condition for participation in the inter-American system. And – just as the IFIs linked democratic governance to the free market – the OAS also advocated economic liberalization and promoted free trade and liberalized investment in the new regional system. This is not to say that the OAS has not led significant initiatives for defending democracy, including the Inter-American Democratic Charter which provides for the suspension of any member state whose democratically elected government is overthrown by force. The OAS has also established a high-level of credibility in observing and monitoring elections, and in establishing high-level roundtable dialogues (so-called mesas) to restore democratic order to Peru (2000) and Venezuela (2002–2004) (Cooper and Legler 2005). It has also entrenched a very powerful anti-coup norm across the region (Legler 2011). Despite important democratic features of hemispheric governance institutionalized by the OAS, however, the organization has been a strong supporter of neoliberal trade and investment policies such as those associated with the failed-Free Trade Agreement of the Americas (FTAA) (Shamsie 2004).

The third major shift was the emergence of Canada as an important force in support of the new emerging regional order. As Gordon (2010) has argued, Canada emerged in the late 1980s as a sub super-power in its own right whose interests were intertwined with those of the United States but which also began to act independently to advance the interests of Canadian capital.[10] Although it supports neoliberal order at the global level, its junior leadership role has been most apparent in the Americas. Canada's growing involvement in the region was spearheaded by business groups such as the BCNI (now the Canadian Council of Chief Executives), which supported the new neoliberal global regime of accumulation (Carroll 2003). The initial strategy focused on continental economic integration, the milestones of which were the Canada–United States Free Trade Agreement (CUSFTA) in 1989 and the NAFTA in 1994. In 1990, Canada finally entered the OAS and became one of the major proponents of the FTAA (the next chapter will discuss the bilateral and plurilateral agreements that have substituted for the FTAA as it failed to gain hemispheric support).

---

10   Gordon (2010) sites the research of geographer Bill Burgess to demonstrate that, although Canadian companies have become increasingly transnationalized, there is a trend of strong Canadian ownership patterns within and across economic sectors. When analysing Canada's role in promoting neoliberalism in the Third World, therefore, explaining Canadian actions exclusively in terms of its subordination to the United States risks obfuscating its own material interests.

Canadian mining companies became particularly influential proponents of trade and investment liberalization. Their share of the larger company exploration market in Latin America and the Caribbean grew steadily beginning in the early 1990s, reaching 35% by 2004. Canadian companies gained the largest share of all of their competitors, with seven companies placing among the top 20 mineral exploration investors in the region from 1989 to 2001. As the mining presence expanded, the Canadian state began aggressively promoting a strategy of 'accumulation by dispossession' robbing indigenous peoples of their land and resources (Webber and Gordon 2008).[11]

The Canadian state also began to take on a more active political role in the region, providing a specific ideological value-added to the new neoliberal regional order. By harnessing its historic reputation as a middle and non-colonial power, Canadian diplomacy was well positioned to confer legitimacy upon the norms and rules of the emerging system. Canada as middle power has supported hegemonic global order by facilitating and mediating and defusing potential destabilizing conflicts, and by sacrificing short-term national interests for the greater good. This role has strengthened the hegemony of US-led regional (and global) capitalism by reinforcing the impression that the order is not narrowly American but based on the common good.[12]

Thus, Canada became an important proponent of the new norms institutionalized by the OAS, using its status as a middle power to promote both neoliberalism and polyarchy throughout the region. In Guyana in the late 1980s, Canada led a donor support group that spearheaded a structural adjustment programme that was subsequently adopted by the government under much pressure (Burdette 1994). In Haiti, it contributed significantly to peacekeeping efforts after President Jean-Bertrand Aristide was restored by the United States in 1994 on condition that he implement a neoliberal programme that prevented him from pursuing any of the reforms for which he was originally elected (this occurred through the Emergency Economic Recovery Program, formulated by the World Bank, IMF, Inter-American Development Bank and USAID with the support of CIDA and several other donors (United Nations 1995)). In Colombia, CIDA played an instrumental role in catalyzing and shaping the revision of Colombia's controversial mining code in 2001, which retracted indigenous and workers' rights, environmental protections,

---

11 Gordon (2010) argues that the strategy of accumulation in the global south is an extension of the centuries-old policy of depriving Aboriginal people of their land and resources at home. The point is of contemporary relevance given that Canadian mining companies are driving this process both in Canada and abroad, and have faced growing resistance on both fronts.

12 The concept of Canada as a middle power emerged in the post-war era as a regulative ideal directing the Canadian state to play an important role in multilateral fora as facilitator and mediator to defuse potential conflicts. Neufeld (1995) follows Cox in using the term middlepowermanship in a neo-Gramscian sense to describe how the concept of middle power was framed in terms of dominant class interests in Canada to reinforce the notion that the global order was not narrowly American, but one which truly represented the common interest.

significantly decreased royalties, and paved the way for privatization (Kuyek 2006). When the authoritarian Fujimori government in Peru – an important ally of both the governments of Jean Chrétien and Bill Clinton – suffered a crisis of legitimacy as the century came to a close, Canada played a lead role brokering the terms of a transition. There, too, Canada and the United States sought to ensure that the neoliberal reforms decreed by the Fujimori government were left intact even as the regime itself lost the ability to govern effectively. Over the next decade, both would sign new free trade agreements and would support a regulatory regime in the mining sector that has led to growing social conflict (see Chapter 4).

At the hemispheric level, Canada became the most vocal advocate of the FTAA – an agreement which would have locked in unpopular neoliberal reforms whose disappointing results were already becoming well known. The government of Jean Chrétien negotiated the early stages of the agreement, chaired an implementation review group and hosted the Summit of the Americas in Quebec City in 2001 to move it forward. With the Chrétien government also championing the Inter-American Democratic Charter alongside the FTAA at the summit, Ottawa's foreign policy in the region was manifestly marked by the same contradictions as Washington's.

Democracy Promotion and Democracy Assistance in the New Regional Order

Patterns of US Interventionism

Such then was the background to the emergence of the current regional order of the Americas. Democracy promotion – in the broad sense of the political and diplomatic positions supported by the United States and Canada in response to developments in the region – was closely tied to an economic project, and the inability of Latin American elites to continue ruling through policies of coercion. Yet despite their shared affinity for neoliberal polyarchy, the United States and Canada have developed very distinct approaches to democracy assistance that reflect the fields of practice which have evolved in each country. For the United States, its approach to democracy assistance was carefully aligned with its overall foreign policy objectives in the region, reflecting the old patterns of stabilization and destabilization; for Canada, this was not the case. Often, Canadian democracy assistance was guided by a grassroots approach to democracy assistance that at times contradicted its commercial interests.

The work of William Robinson has traced the emergence of US democracy promotion to policy discussions in the late 1970s and early 1980s when US elites earnestly began debating the most appropriate political model for achieving social control and stability in Third World countries.[13] Organic intellectuals and state managers associated with both the US state and the emerging transnational elite reflected upon the structural changes to reorient policy, advocating the use of an underdeveloped and underutilized instrument in US foreign policy typically associated with covert CIA operations – political aid – to guide the democratic transitions in a direction that would ensure the stability in the interests of the emerging global capitalist economy.

The new democracy assistance agencies which launched the first wave of democracy wars were located within and outside the state. From within, USAID's Office of Democratic Initiatives (ODI) was created in 1984, before the Office of Transition Initiatives (OTI) took over its functions in 1994. The OTI targets short term political assistance to countries undergoing political transitions or experiencing political crises. US Ambassadors play a key role in this infrastructure, strategically orienting democracy programmes to accomplish foreign policy objectives. Since 2001, USAID's Office of Democracy and Governance (ODG) has coordinated US programming in these areas in collaboration with local USAID missions.

Outside the state, the NED and its sister organizations were established in 1983 by an act of Congress. The NED affiliated organizations are the International Republican Institute (IRI), the National Democratic Institute (NDI), the Center for International Private Enterprise (CIPE), and the American Center for International Labor Solidarity (Solidarity International). From its inception, the NED was closely integrated into official policy circles in Washington, though its credibility and that of its sister organizations have always been based on their highly questionable non-governmental status (often referred to as quasi non-governmental organizations, or QUANGOs). As Guilhot (2005) argues, the democracy promotion field of practice is in fact defined by a close network of state agencies linked together by 'double agents' who move back and forth freely between them. Over the years, the endowment's board of directors has regularly included individuals who have served at the highest levels of the foreign policy establishment, as has been the case with both IRI and NDI. Endowment board members have included former US secretaries of state Henry Kissinger and Madeleine Albright (who is also the current chairman of the NDI), former US Secretary of Defense Frank Carlucci, former national security advisor Zbigniew Brzezinski, a former NATO supreme allied command in Europe, General Wesley K. Clark, and the former deputy secretary of defence and head of the World Bank, Paul Wolfowitz. The current chairman of IRI is John McCain. In addition to the QUANGOs, there are also a multitude of private contractors who work with USAID to implement democracy assistance programmes such as Chemonics International and ARD Inc. according to the needs and objectives of their state funders.

Despite the explicit links to the US government, foreign policymakers and intellectuals have repeatedly depicted the endowment and its sister organizations

---

13   See Robinson (1996) for a rich analysis of the key policy documents and decision makers who articulated this emerging strategy.

as independent and neutral NGOs. As a grant-giving institution, there is no question that the endowment has supported civil society organizations and political parties of many ideological stripes. US democracy promotion agencies as a whole implement hundreds of programmes across the globe and undoubtedly provide support to actors from across the political spectrum. NDI in particular has adopted a fairly pluralistic approach to its democracy promotion programmes, maintaining formal ties with the liberal, socialist, and centrist democrat internationals.

Since its inception, however, the endowment's approach to pluralism in some countries has been compromised by its intense ideological biases in others. Although its opposition to specific governments is invariably justified in terms of principled concerns with the quality of democracy, countries that are singled out tend to be those that oppose US interests (as we will see in the next chapter). This reflects how democracy assistance has always been embedded within foreign policy strategies that have very little to do with supporting democracy. Neo-conservative ideology was particularly important in shaping the nascent democracy promotion industry in the context of the Cold War under the Reagan administration. When the endowment was established, democracy promotion was used to guide transitions from authoritarianism away from socialist alternatives while bypassing stable authoritarian regimes that remained allies of the United States (and which, under the Kirkpatrick doctrine, were deemed more likely to eventually democratize than 'totalitarian' socialist regimes). Democracy promotion was a key component of the government's commitment to low-intensity conflict to rollback socialist threats.

Although many Democrats supported the new form of intervention, others led efforts in the House of Representatives to cut off funding for the NED in 1993. The following year, however, funding was actually restored, with an increase from $30 million to $35 million (Rieffer and Mercer 2005). Under the Clinton administration's grand strategy of 'democratic enlargement', democracy promotion was discursively linked to the prosperity of the 'new world order' and the new commitment to free trade. The liberal-democratic triumphalism of the 1990s no doubt contributed to a more favourable climate for the NED and other US agencies, which managed to avoid major controversies as democracy promotion became a truly bipartisan affair.

Empirically, critical researchers have shown how US democracy promotion has undermined popular movements or governments throughout much of the Americas. Nicaragua and Haiti in the late 1980s and early 1990s were the early laboratories in which the new form of political aid was tested (Robinson 1996), though more recent research has shown how US democracy assistance has operated as a form of interventionism in Venezuela (Allard and Golinger 2009, Barry, 2007, Cole 2007, Clement 2005), Bolivia (Lindsay 2005, Beeton 2009, Dangl 2008, and Bigwood 2008) and Honduras (Golinger 2009).

In theoretical terms, democracy promotion mobilizes networks and coalitions in civil society to wage what Gramsci referred to as the war of position – a form of ideological trench warfare in the struggle for hegemony – against left and centre-left governments. As a whole, US agencies and the aid that they provide 'coalesce into a complex and multilevel US intervention network' (Robinson 2006: 106).

US programmes are of course also implemented in countries where allies are in power. The long-time head of the National Endowment for Democracy (NED), Carl Gershman, has distinguished between two types of democracy promotion efforts – those fostering 'long-term democratic political development' and those aimed at securing a 'democratic transition' or change of regime (quoted in Robinson, 1996). In Mexico's fragile neoliberal polyarchy during the presidency of Ernesto Zedillo (1994–2000), democracy promotion was more about stabilization than regime change (which, as we shall see, has also been the case in Peru and, at times, Bolivia). According to the neo-Gramscian theorist Adam David Morton, US programmes reinforced the hegemony of neoliberal polyarchy through support for moderate NGOs and human rights organizations adapting civil society activism to institutionalized liberal democracy. This more systemic orientation reflected the Mexican state's own ability to dissipate class struggles through engineered social and political reform – a strategy of passive revolution that co-opted popular demands without substantially expanding mass control over politics.

As a form of hegemony linked to the global economy, Robinson argues that US democracy promotion benefits transnational capital as a whole. But as a foreign policy practice, democracy promotion stems from the US historical pattern of imperial intervention and unique set of ideological justifications stemming from its position in the world system (Grandin 2007). In this sense, it reflects the duality of the *raison d'état* of the US state, which seeks to organize production both nationally and internationally in accordance with the interests of transnational capital while pursuing American geopolitical goals in the interstate context.[14] The US democracy assistance field of practice, in short, has evolved in a unique national-ideological setting defined by a close network of state agencies and NGOs which have always been linked to a very undemocratic foreign policy.

Canada's Less Instrumental Approach

The evolution of the Canadian democracy assistance field of practice occurred in very different circumstances. From a critical perspective, it may be viewed as the product of the tensions between the humane internationalist (Pratt 1990) strand of Canadian development and the increased subordination of development programming to neoliberal foreign policy objectives. Humane internationalism has historically been linked to Canada's history as a non-imperial power, fairly progressive NGOs, a socially-active Church, and social democratic values. At

14 This approach also resonates with Panitch and Gindin's (2003) analysis of US imperialism, which argues that while the US state may serve as a 'general coordinator for international capitalism' – a metaphor they borrow from Perry Anderson – it continues to use its power to advance and protect interests defined in national terms.

one point, this conception helped inform a more progressive notion of Canada's role as a middle power (Pratt 2001). While there have always been assumptions of cultural superiority that have historically underpinned Canada's approach to development (Macdonald 1995), the concept helps illuminate the different historical and institutional circumstances in which the Canadian democracy promotion field of practice developed as a subfield of international development. Figure 1.1 below highlights the key differences in the US and Canadian fields of practice according to the actors involved, the networks they are part of at home, whom they support abroad, and the different visions of democracy with which they are associated.

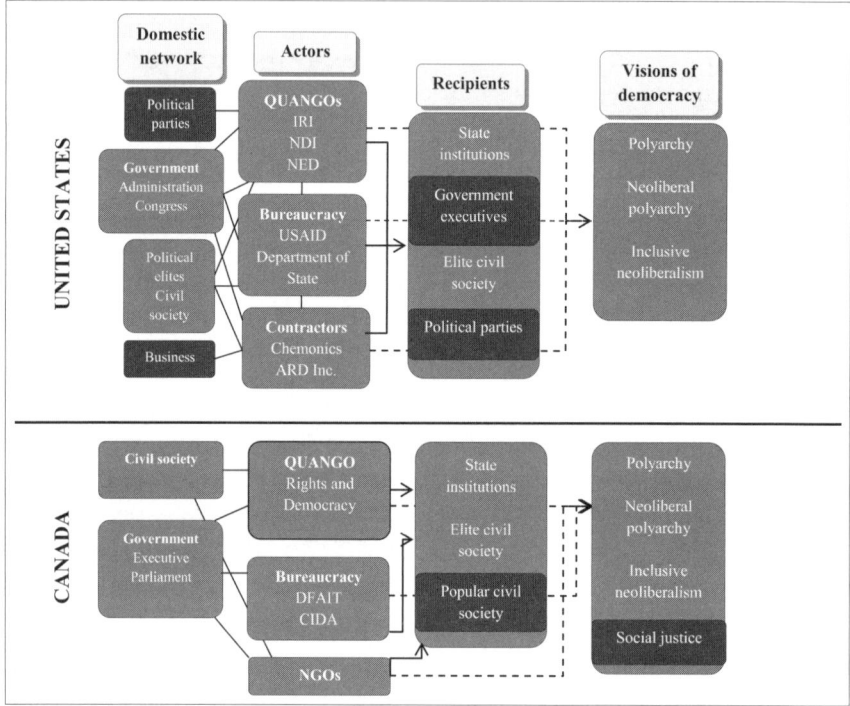

Figure 1.1 US and Canadian fields of practice in democracy assistance
Note: General distinctions between the US and Canadian fields of practice appear in dark grey (though exceptions can be found). The list of actors is by no means exhaustive, but includes those which have been most prominent historically (note that Rights and Democracy was recently closed).

The bulk of Canadian democracy assistance has traditionally been channelled through the Canadian International Development Agency (CIDA), which supports democratic development through a wide range of Canadian and international NGOs, national governments and multilateral institutions. Many of the organizations involved in Canadian democracy promotion focus on strengthening specific institutions (for example, the Parliamentary Centre). But there also has been a significant focus on supporting civil society, particularly through Rights and Democracy. Originally named the International Centre for Human Rights and Democratic Development, Rights and Democracy was mandated by the Canadian parliament in 1988 to support the universal values of human rights and the promotion of democratic institutions and practices.

Unlike the National Endowment for Democracy, the organization was not conceived of as an instrument to advance foreign policy objectives in clearly defined ideological terms (although Canadian policymakers did look to the endowment and German political party foundations as potential models). It was not associated with a government policy of democratic interventionism but rather with Canada's growing development efforts. Whereas the endowment was led by neoconservatives close to the Reagan administration, Rights was launched by the more moderate Progressive Conservative party; its first president, moreover, was a committed social democrat (former NDP leader Ed Broadbent).

The special joint committee that called for the establishment of the agency recommended an inclusive approach to human rights, eschewing the very term democracy promotion because of its association with the philosophy of the 'present USA administration' (Schmitz 2004). The concern for differentiating Canadian activities in the area of democracy promotion has carried over until very recently. According to one CIDA official, Canada has traditionally supported democratic development – not promoted it – since the emphasis has always been placed on strengthening democracy in collaboration with governments that have requested support.[15] Rights and Democracy's approach emphasizes citizen participation in the public sphere as both a right and a *sine qua non* of democratic consolidation (Thede 2002). One of the few comparative studies on the NED and Rights and Democracy found that the latter has focused almost exclusively on supporting human rights groups while the former has provided considerable funding strengthening business or pro-market civil society organizations. The board of Rights was also historically composed mainly of international and national human rights activists and academics rather than high-ranking foreign policymakers (Scott and Walters 2000).

In many cases, moreover, Rights and Democracy has taken a highly critical approach of Canadian foreign policy, as well as the activities of Canadian multinationals. The same holds true for the various NGO involved in democracy promotion that receive much of their funding from CIDA and Foreign Affairs. Development and Peace, for instance – the official international development organization of the Catholic Church in Canada and one of the most important contributors to democratic development – supports grassroots civil society groups around the world from a liberation theology perspective. It has publicly criticized

---

15   Interview, Gatineau: 19 December 2009.

many of the neoliberal policies that have led to increased poverty and inequality (as we shall see in Chapter 4, it has also denounced the practices of Canadian mining companies).[16] The Canadian Council for International Co-operation, a coalition of NGOs, includes many members who have sought to balance their commitment to issues of social justice with their need for state funds to implement programmes. Some, like the NGO Inter Pares, have worked in solidarity with popular movements in Central America since the 1980s.

CIDA, for its part, has traditionally focused its democracy assistance on institution building rather than empowering civil society groups or political parties. It has supported polyarchy, but not – at least historically – at the expense of other visions of democracy. The agency articulated a progressive commitment to a more expansive, rights-based notion of democracy in its 1996 policy, Human Rights, Democratization and Good Governance, which emphasized the importance of empowering citizens through its democratization programmes. This commitment also resonated with the concept of human security championed by then-Foreign Affairs Minister Lloyd Axworthy in the late 1990s.

At the same time, there are longstanding tensions in CIDA's approach to supporting good governance, which has been linked to structural adjustment policies since the late 1980s (as noted above). The agency's commitment to the humane internationalist strand of Canadian foreign policy has steadily eroded over the years, giving way to the gradual triumph of commercial and geopolitical interests in the formulation of aid policy (discussed in depth in the following chapter).

Political Objectives and Hegemonic Strategies of Democracy Assistance

Before we turn to contemporary developments in the regional order and the evolution of US and Canadian democracy promotion in the new millennium, the following provides a brief overview of the political objectives that democracy promoters have historically sought to accomplish and the strategies and tactics they have used (summarized in Table 1.1). The typology is based on the preceding discussion as well as the arguments that will be developed more fully in the context of the case studies.

Democracy assistance is inherently political insofar as it alters (or has the potential to alter) the balance of power between different social forces. Approaches to democracy promotion may be distinguished, however, based on their political objectives, who they benefit, their strategies, tactics, how they affects hegemonic relations, and the form of democracy they ultimately promote. From this perspective, we can distinguish between three different

---

16   At the time of the FTAA discussions, for instance, Rights and Democracy published a report which was highly critical of the track records of previous trade agreements Bronson and Lamarche (2001).

democracy promotion strategies related to democracy assistance: 1) those that destabilize left-nationalists and strengthen neoliberal and elite social forces; 2) those that stabilize the social order; and 3) those that empower grassroots movements.

Table 1.1 Political approaches to democracy assistance

Political objectives	Hegemonic strategies	Tactics and practices
Destabilization (regime change)	War of position/crisis of authority	Mobilize elites and facilitate coalition building in political and civil society; boycott left government or popular party; legitimize democratic interventionism at home
Stabilization	Passive revolution	Build legitimacy of government; contribute to process of state reform; strengthen democratic institutions; support mainstream NGOs; redefine citizenship according to liberal norms
Grassroots empowerment	Counter-hegemony	Support and accompany grassroots organizations and progressive NGOs based on solidarity and accompaniment

Destabilization or regime change is characterized by the mobilization of elite coalitions in political and civil society to engage in a war of position against a popular government or movement. When the target is a government, the opposition seeks to generate a crisis of authority for the state, often by using superior access to communications technology to advance claims of human rights abuses. Indeed, human rights have become a key battleground on which the war of position between competing social forces is often waged, with both sides seeking to capture or maintain popular legitimacy. The war of position is also carried out at home on North American soil by politicians and democracy promoters alike seeking to influence public opinion or the policy process.

Stabilization occurs when the focus of democracy assistance is to consolidate a neoliberal polyarchy that faces instability and social conflict. A stabilization approach typically includes two basic dimensions: 1) the strengthening of traditional democratic institutions and new mechanisms of inclusive neoliberalism that help manage social conflict; and 2) an attempt to shape the public discourse on democracy and create new liberal subjects in civil society.[17] Stabilization tactics

---

17 I borrow the term from Craig and Porter (2006), who argue that inclusive neoliberalism adds positive liberal approaches to emphasizing empowerment to enable

focusing on the state include strengthening legislatures, reinforcing electoral processes, and sponsoring or supporting laws, bills, policies, codes and reforms. Participatory initiatives, such as decentralization schemes that facilitate local citizen engagement without addressing redistribution issues, tend to be a favourite. Ideologically, the stabilization approach resonates with a post-Washington consensus concern with broadening participation and social inclusion without challenging the actual distribution of wealth and power. The focus is primarily on managing citizens' expectations and changing popular perceptions to demonstrate that the system does in fact work. As such, they reinforce the state's project of passive revolution.

In civil society, stabilization tactics support organizations rooted in the urban professional and middle class sectors, some of which are uncritical of neoliberalism but others which may be led by left intellectuals with more activist backgrounds. While some may be inspired by a radical philosophy of social change, they tend to be pragmatic and moderate in their demands, with little connection to the social movements engaged in the day-to-day struggles for social existence. Instead, they focus on thematic issues and depend upon the international community for funds to implement specific projects. In a context of polarization, they may side with local elites over popular movements, as we will see in the case of Haiti. In short, they tend to be part of elite as opposed to popular civil society (see Table 1.2 below). Those grassroots NGOs with a more substantial social base that do receive support often become professionalized and incorporated into mainstream civil society networks. In this sense, such NGOs project a form of capillary power (Morton 2007) as institutions that have penetrated much of civil society and which transmit hegemonic discourses that contain demands for change within a liberal democratic framework. They serve a particular function of socializing subaltern groups into liberal citizenship norms and de-politicizing them, thereby contributing to the state's project of passive revolution.[18] In preaching the gospel of social compromise even as they promote reforms, they place safe limits on structural transformation.

---

participation to core conservative neoliberal macroeconomic and pro-market policy settings.

18   Although the approach of this book is grounded in historical materialism, some readers will note the use of Foucauldian concepts such as discourse. In using such terms, I take inspiration from the cultural political economy approach of Sum (2005), who integrates micro concepts of power to describe how hegemony is produced within world order.

Table 1.2 Elite versus popular civil society

Elite civil society	Led by political elites, dominant class, middle class, and dominant ethnic groups; urban based Includes: business and private sector associations, conservative think tanks, TNCs, media, church groups, consultancy firms, philanthropic organizations
NGOs	Based in elite civil society but ideologically diverse; represent subordinate groups rather than organize and mobilize them; rely on donor funding; radical NGOs may bridge the gap with popular civil society
Popular civil society	Rooted in subordinate classes and social groups; institutionally weak Includes: popular social movements, trade unions, grassroots and popular organizations, indigenous movements; radical dioceses and church groups

But there are differences in the ways Canada and the United States go about stabilization. As we will see, the US approach tends to be more partisan insofar as its programmes seek to build the legitimacy of specific governments and strengthen their political-ideological projects. Canada's approach focuses more on building institutions than enhancing the legitimacy of governments. This is also the case with some US actors, such as NDI. The inter-subjective understandings of US and Canadian actors of what they are accomplishing also varies considerably – Canadian officials are much less likely to view their activities as political, even though structurally speaking they contribute to a process of stabilization which is anything but politically neutral. Viewed within the larger pattern of foreign policy engagement, building liberal-democratic-institutions supports neoliberal polyarchy when coupled with neoliberal economic policies.

For both Canada and the United States, those who promote polyarchy typically operate on the assumption that the foreign policy interests of their home states are benign, or are at least reluctant to criticize them (especially when they are dependent upon state funds). The lack of 'democratic development' is conceptualized mainly in terms of cultural factors rather than the complex historical relationships of domination between states and social classes. In this sense, democracy promotion is inscribed in a particular epistemic community characterized by a lack of critical thinking. Questions on the political interests behind democracy promotion are rarely asked, and most actors are hamstrung by the latest development fads articulated by their funders. Ultimately, democracy promotion is saturated by discourses of cultural superiority in which western states get to determine what democracy means (Dhaliwal 1996), whether or not specific democracy assistance initiatives are consciously mobilized in support of a political-ideological project.

In terms of promoting more substantive visions of democracy, many Canadian NGOs do support grassroots organizations that maintain a popular social base and commitment to transformative action. Macdonald's (1994) research on NGOs in Central America identifies key concepts to help conceptualize the approach, which is based on accompaniment, solidarity and respect for control by the local partner within a larger commitment to the process of social change. Activities may include technical training, brigades, exchanges and human rights advocacy, as well as financial support. Such activities may also contribute to a process of consciousness-raising surrounding the social struggles of partner organizations. They contribute to counter-hegemonic constructions of democracy that seek to extend popular control over parts of the economy, emphasizing structural transformation and social change.

At the same time, however, the grassroots approach remains embedded in unequal relations between North and South; ultimately, Canadian NGOs get to determine who receives support in local political struggles with high political stakes. As I will argue in Chapter 3 on Haiti, they can misread the political situation based on weak evidence, ethnocentrically and arrogantly dismissing arguments that call into question their activities. Furthermore, since many democracy promoters claim to be contributing to grassroots organizations – including even the most conservative ones – it can be difficult to differentiate between a genuine grassroots approach and a more subtle strategy of co-optation.

Concluding Remarks

A critical analysis of the historical evolution of US and Canadian democracy promotion in the Americas reveals the contradictions, tensions, and paradoxes of this complex field of international development. When democracy promotion is situated within the larger foreign policy interests and objectives that characterize relationships between states – that is, when it is looked at as an enactment of power with undemocratic political implications – the notions of altruism that shroud this area of intervention evaporate. For democracy promotion in the general sense has historically been linked to an economic project that is fundamentally undemocratic in nature, in terms of how this project has been carried out, and in terms of its impact on people's lives and their ability to participate in democracy.

In the Americas, both the United States and Canada have used democracy promotion as a modality of power intended to secure a hegemonic vision of regional order predicated on neoliberal polyarchy. For the United States, moreover, the commitment to this limited form of democracy only emerged as authoritarian regimes lost the capacity to govern effectively, and as new political arrangements were required to stabilize the emerging order. Democracy assistance programmes therefore evolved to respond to the shifting balance of power in specific countries. At the same time, the field of practice delivering Canadian democracy assistance has historically been more autonomous from the Canadian state's economic project in the region, thus indicating the need to take into consideration how national institutions, identities, and interests have shaped the different approaches of Canada and the United States.

Chapter 2
North American Democracy Promotion in the New Conjuncture

With US global power on the wane, Washington's ability to intervene against its adversaries in Latin America as it did throughout much of the twentieth century has been significantly curtailed. The rise of regional powers like Brazil, the proliferation of new regional bodies outside the scope of US influence, and widespread resistance to neoliberal globalization and regionalization in many countries has placed new limits on US power. Yet, the United States continues to exert considerable influence on the domestic politics of the region through a range of foreign policy practices, including democracy promotion. In the quest to deepen and strengthen neoliberalism, Canada remains an important ally, and one which has increasingly adopted a more strategic approach in situating its own democracy promotion efforts within the shared North American vision of regional order. Latin America remains a region of vital interest to both Ottawa and Washington, and the foreign policies of both countries have intersected with the shifts in regional power and the cycle of political radicalization that marks the current conjuncture.

This chapter will explore how Canada and the United States have responded to these important political changes. Its purpose is to situate developments in democracy promotion in relation to larger commercial and geopolitical interests, and to identify the key themes that will be explored in greater depth in the case studies. The chapter is divided into three parts. I begin by looking at the political changes that have swept Latin America in the last decade, including the characteristics of both the new left regimes and conservative holdouts, the drive towards regional integration, and the ongoing importance of the global economy. I conclude by briefly examining commercial relations between Canada and the United States and the rest of the region.

The remaining sections then examine the regional foreign policies and democracy promotion strategies of the United States and Canada chronologically, beginning with the Republican administration of George W. Bush (2001–2009) and the Liberal governments of Jean Chrétien (2000–2003) and Paul Martin (2003–2006). During this period, the Bush administration sought to build civil society oppositions to left governments in Haiti, Venezuela, and Bolivia. At the same time, those states which remained in the US sphere of influence were supported through militarily aid and democracy assistance programmes designed to prevent 'anti-systemic' movements from gaining further political ground. The War on Terror and the War on Drugs provided further fodder to rationalize strategies of stabilization and intervention in the region. For the first time, however, US efforts at regime change led to a backlash against democracy promotion, with the governments of Venezuela and Bolivia at the forefront of regional opposition. This led many liberal internationalists to call for a softer approach and to insist upon the necessity of situating US activities within the larger international democracy promotion effort. The Bush administration also sought unsuccessfully to reconstitute regional hegemony by softening the economic discourse of the post-Washington Consensus without actually reorienting policy.

Canada's engagement in the Americas during this period was marked above all by its policy in Haiti. Under the Liberal government of Paul Martin in the early 2000s, Canada articulated a 'Responsibility to Protect' doctrine that helped legitimize the transnational campaign against Jean-Bertrand Aristide. Although the Liberals did not specifically undermine the rights-based tradition discussed in the previous chapter, democracy promotion as a whole became increasingly subordinated to security and commercial interests as Canadian middlepowermanship took on a much more aggressive form. Canada also prioritized other countries in the region where it held growing mining interests – such as Guatemala and Peru – for democracy assistance.

The third and final section examines the regional foreign policy and democracy promotion records of the Democratic administration of Barack Obama (2009–), and the Conservative government of Stephen Harper (2006–). During this period, the Obama administration retained the commitment to deepening and strengthening neoliberalism while adapting to the new regional constraints on US power through a more cordial 'good neighbor imperialism' (Vanderbush 2011). Although the NED continued to undermine the governments of Cuba and Venezuela, democracy assistance became even more focused on propping up security-state polyarchies such as Colombia, Mexico, and Haiti as stabilization and securitization became the twin strategies of reproducing regional hegemony. The escalating War on Drugs under Obama's watch provided human-rights offenders like Colombia and Mexico with massive military resources as coercion increasingly substituted for consensus. The implicit support that coup forces in Honduras received by the United States (and Canada) further undermined any hope that Obama would break with the policies of his predecessor.

Under the Harper government, Canada's 're-engagement' with the Americas has also been marked by continuity with the previous government with security and commercial objectives at the forefront. Like the Obama administration, the Harper government has prioritized relations with security-state polyarchies while adopting an increasingly hostile approach towards Venezuela. Unlike the United States, however, it has maintained positive relations with Bolivia. The Harper government has also sought to leverage Canada's reputation as a more social democratic nation to reframe its own hegemonic aspirations. Underpinning these macro shifts is a concerted effort on the part of the Harper government to replace the Canadian development field of practice with more compliant networks (for example, those willing to work with Canadian mining companies), a development which threatens what remains of Canada's rights-based approach. As the first decade of the new millennium drew to a close, Canada and the United States were increasingly acting in concert to enforce neoliberal regional order within an ever smaller sphere of influence.

Threats to North American-Led Neoliberal Regional Order in the New Millennium

Hugo Chávez's victory in Venezuela's 1998 presidential elections marked the beginning of the 'pink tide' which swept left governments to power throughout Latin America. Over the course of the next decade, left and centre-left governments were elected in a dozen countries on the basis of opposition to the neoliberal policies of the Washington Consensus. Scholars of different approaches agree that the origins of the left electoral victories reside in the crisis of the neoliberal state and the mass resistance engendered by it as social movements organized to contest the policies of inequality and exclusion (Petras and Veltmeyer 2009; Cameron and Hershberg 2010; Silva 2009). For while the political elites who managed to impose the neoliberal project upon the state in the wake of the democratic transitions came to exercise hegemony in political society – forcing even leftist parties to except the strictures of global capitalism – they failed to achieve hegemony in civil society. The contradictions of global neoliberalism, including increased poverty, pauperization, and social precariousness, fuelled the explosion of anti-neoliberal popular movements (Robinson 2008). Although the new Latin American left does not present a standardized discourse or explicit set of practices, it has emerged as a 'collection of principles, ideas, practices and institutions' (Dagnino, Olvera and Panfichi 2008: 27).

Attempting to categorize the new regimes, however, invariably stirs up controversy. If liberal and conservative intellectuals typically deploy some variation of Jorge Castañeda's dichotomy to distinguish between a responsible social-democratic left committed to the market (Brazil, Chile and Argentina) and a supposedly irresponsible one seeking socialist transformation (Venezuela, Bolivia, and Ecuador), more radical scholars often suppress important differences in the name of solidarity (for example, French 2010). But to understand the political consequences of democracy promotion, we must also consider the nature of the different hegemonic projects being articulated by left and right alike. In this respect, Gramsci's continuum of coercion and consent, summarized in Table 2.1, provides a useful framework.

None of the Latin American states have managed to achieve an integral hegemony, although Venezuela and – to a lesser extent – Bolivia and Ecuador, have challenged class relations and have sought to articulate new hegemonic ideologies that have mobilized popular forces. Both 'twenty-first century socialism' in Venezuela and Ecuador and indigenous socialism in Bolivia have supported the use of referenda to enable mass participation in key decision making, recall (*revocatoria*) to ensure ongoing accountability of elected officials, and participatory institutions at the local level. Venezuela's horizontal and vertical councils have provided the most important institutional space to both mobilize and facilitate mass political participation. In Bolivia, the democratic experiment has also recognized traditional indigenous practices such as communal justice, emphasizing the importance of decolonizing the state, popular sovereignty, redistribution, and social justice. As Patzi (2004) observes, the notion of indigenous democracy in Bolivia is defined by the subordination of authorities to the community; the representative 'leads through obedience' (*manda porque obedece*) to the people.

Table 2.1 Forms of hegemony and class rule

1	*Integral hegemony*	Class compromise between dominant and subordinated social forces
2	*Decadent hegemony*	Hegemonic ideology but institutional mechanisms of compromise are weak.
3	*Minimalist hegemony*	Ideological unity only among intellectual, political and economic elites strongly averse to any form of intervention by popular social groups in state life. Strategy of cooptation (*trasformismo*) prevails and the state resorts to a politics of passive revolution.
4	*Class supremacy*	Ruling class dispenses with illusion of consent and dominates society through organized violence and coercion.
5	*Caesarism*	Dominant and subordinate classes balance each other in force; ongoing struggle between them would lead to their mutual destruction. A strongman forces a compromise to break the class stalemate.

But both remain polarized between dominant and subordinate classes, with the traditional elites retaining much of their economic power. Their democratic experiments are fragile; at best they have achieved a decadent form of hegemony. The strong cult of personality in both countries, moreover, is as much of a threat to the new vision of participatory democracy as it was to previous socialist projects (see López Maya (2008) for a left critique of *chavismo* in Venezuela). In Bolivia, some have warned that indigenous democracy often reproduces gender inequality and contains certain authoritarian tendencies in its own right, including the use of violence to intimidate political opponents (Thede and de la Fuente 2008). More radical scholars such as Petras and Veltmeyer (2009) and Webber (2011a) have argued that Bolivia has deepened a dependent-structure of accumulation in the world capitalist economy that privileges the interests of the agro-mineral oligarchy over the peasantry and urban working class. Along with most other left and centre-left governments in the region, it has pursued the modest social policy of the post-Washington Consensus without altering the underlying structure of the neoliberal economy. The commodities boom of the 2000s has provided the shallow foundations for the new approach.

Although the more radical governments have not broken from neoliberalism, they do have the critical support of the social movements, which are trying to recapture the process of change (Robinson 2011). Bolivia, Ecuador and Venezuela have also taken an unequivocal stance against US interventionism and have led a sub-grouping of left nationalists through the *Alianza Bolivariana para los Pueblos de Nuestra América* (Bolivarian Alliance for the Americas – ALBA). While most trade flows within the alliance are with Venezuela – which provides member states with oil at discounted prices – its principles provide an alternative to the ideology of free trade. Each country has also asserted its diplomatic independence by building closer relations with China, Iran and Russia (as has Brazil). As we will see, US policymakers and programme officers have been very clear about their antipathy towards these governments, which are denounced as 'populist' and 'anti-systemic'.

Brazil, Argentina, and Chile are also considered to be part of the new regional left. These countries have moved towards a decadent form of hegemony ensuring some degree of stability and social peace based on redistributive measure (such as the *Bolsa Familia* in Brazil) that ultimately leave the power of dominant classes untouched. The Brazilian sociologist, Francisco de Oliveira, has dubbed the political situation in Brazil a form of inverted hegemony, whereby the political leadership of the Workers' Party is recognized (and in some cases, even supported) by the dominant classes, but the party itself has achieved the consent of the dominated to the structures of their exploitation (Anderson 2011). At the same time, the moderate left has also rejected US intervention in the region and has played an important role in Latin American efforts to offset traditional US dominance through initiatives to promote greater political and economic integration, such as the *Unión de Naciones Suramericanas* (Union of South American Nations – UNASUR), an intergovernmental union that will integrate Mercosur and the *Comunidad Andina* within an EU-like structure, and CELAC, which, as previously noted, includes all sovereign countries in the Americas except Canada and the United States.

Brazilian capital has also expanded across the Americas, and Brazil has become a sub-imperial power in its own right (Flynn 2010). Its ascendancy coupled with the inability of the United States to keep in check openly hostile governments signify profound changes in both regional and world order. Both the symptoms and causes of US relative decline include imperial over-extension, the rise of the BRIC quartet (formed in 2009), the hollowing-out of the US economy, financial collapse and the Great Recession.

The new opposition to US dominance, however, has not led to a break with global neoliberalism; on the contrary, each of these countries has embraced the global economy as part of their accumulation strategy. They have rejected, however, North American-led regionalism in favour of deepening their ties with new trading partners like China, which is now the region's third largest trading partner behind the United States and the European Union (which it will soon overcome). China is currently Brazil's top commercial partner, with US$ 56.4 billion in trade between the two countries in 2010 (Presidency of the Federative Republic of Brazil 2011). Chinese investment in the region is rapidly expanding; in 2010 alone, China invested over US$ 15 billion, making it the third largest investor in the region behind the United States and the Netherlands (ECLAC 2010b).[1]

Taken together, these developments have thrown the regional order that emerged at the end of the twentieth century into a state of flux and transition. While it is unclear what will emerge from these changes (a topic I will return to in the concluding chapter), the struggles that have occurred indicate that the construction of regional order is an ongoing problematic from the perspective of those who favour a monolithic vision.

The changes in the inter-American system and the rejection of the Washington Consensus by the vast majority of Latin American people – if not their governments – has led to a crisis of authority for the United States. In this period of contestation, both the United States and Canada have increasingly looked to the more repressive regimes of the region to maintain geopolitical and economic influence. This has included security-state polyarchies such as Mexico, Colombia, Peru, Honduras and Haiti. Although democratic institutions in the first three countries are much more consolidated than in Haiti, each of them has contained the fallout of inequality (drugs, insurgency, crime, rebellion and social protest) through military and police repression. They have maintained minimalist forms of hegemony relying heavily on a combination of passive revolution and coercion.[2] In the case of Haiti, a form of caesarism prevails; the refusal of the dominant class, itself highly fractured, to compromise with subordinate classes – and the inability of one force to triumph over the other – has led to a political stalemate broken in favour of the elite only by the presence of an international occupation force.

But the North American economic presence in the hemisphere is still strong. With a combined GDP of US$ 1.7 trillion (IMF 2010, cited by DFAIT 2011b), Latin America and the Caribbean remain a region of vital economic interest to both Canada and the United States. Between 1998 and 2009, total US merchandise

---

1   As Gallagher (2010) warns, however, growing trade with China carries with it significant risks as Chinese manufactures increasingly flood the Latin American market. Latin American countries overwhelmingly export primary commodities. The region's bilateral trade deficit with China skyrocketed from $863 million in 2000 to $32 billion in 2009, with only Brazil, Chile and Peru maintaining trade surpluses with China based on their exports of iron ore, copper and soybeans (Paus 2011).

2   Morton's (2011) historical analysis of passive revolution and state formation in Mexico in the post-revolutionary era provides a recent and compelling application of different forms of hegemony in a Latin American state.

trade (exports plus imports) with Latin America grew by 82%, considerably faster than for any other region with the exception of Africa (where trade growth was based mainly on oil imports) (Hornbeck 2011). Although US trade has decreased in the wake of the Great Recession, total good exports to the Western Hemisphere (excluding Canada) were US$ 236.8 billion in 2009, while imports equalled US$ 284.2 billion.

US foreign direct investment (FDI) stock in the countries of the western hemisphere (including Canada) was US$ 791.1 billion in 2008, up 6.5% from 2007. FDI in the United States from the same countries was US$ 271.1 billion in 2008, up 2.8% from 2007 (Office of the United States Trade Representative 2011). In 2010, the United States was still the main investor in Latin America and the Caribbean, injecting 17% of all FDI, followed by the Netherlands (13%), China (9%), and Canada, Spain and United Kingdom (with 4% each) (ECLAC 2010b). Like most FDI (including Canada's), US investment is concentrated in services which were deregulated and privatized throughout the 1990s, such as telecommunications, financial services, and public utilities, as well as in natural resources (hydrocarbons and metal mining).

Latin America and the Caribbean is currently the most important destination for Canadian FDI, which reached CAD$ 93.4 billion in holdings at the end of 2009 (DFAIT 2011b). Canada is also the third largest investor in South America, following the United States and Spain (ECLAC 2010b). Canadian FDI is concentrated mainly in mining, financial services and some manufacturing. Canadian merchandise exports to the countries of the region reached $4.4 billion in 2009, while imports equalled $10.3 billion. Canada exports vegetable and mineral products as well as some industrial and high-value added goods while importing primary commodities from the region (particularly precious metals, minerals, and vegetable products) (Randall 2010a). Canadian services exports were $2.8 billion in 2007, a significant portion of which was in the form of commercial services provided to the Caribbean, particularly in the financial sector. Services exports increased by 10.3% from 2002 to 2007 (DFAIT 2011b). For both the United States and Canada, these economic interests have played an important role in shaping foreign policy and the response to the leftist shift through a new round of democracy wars.

The New Interventionism: Republicans and Liberals

Bush and the Nationalist Resurgence

The Bush administration set the initial tone for the way the United States would respond to new threats to regional order through democracy promotion in the early 2000s. Although Latin America did not occupy a central position in the administration's foreign policy, which focused on the Middle East, the return of hard right appointees revived the old patterns of interventions. Underpinning these developments was the transformations of democracy promotion as a revolutionary method advancing national security by Bush's post-9-11 National Security Doctrine. The poorly conceived invasion of Iraq was meant to demonstrate US willingness to enhance its security through pre-emptive regime change (referred to as the Bush Doctrine) in a 'crusade' to spread democracy. In Latin America, US opposition to popular governments was justified by a 'securitization' discourse which depicted anti-neoliberal social movements and governments a threat to US and hemispheric security (Carranza 2009).

During both of his administrations, Bush appointed veterans from Reagan's Central American policy to key foreign policy positions, including Elliot Abrams as Deputy National Security Advisor for Global Democracy Strategy, John Negroponte as Deputy Secretary of State and later UN Ambassador, envoy to Iraq, and intelligence czar, and Otto Reich as Assistant Secretary of State for Western Hemisphere Affairs.[3] Negroponte, Reich, and his successor, Roger Noriega, played an important role in radicalizing the administration's regional foreign policy (Emerson 2010).

USAID's budget for democracy programmes steadily increased (as did that of the National Endowment for Democracy's), from $1.2 billion in 2004 to $1.7 billion in 2008, at which point it accounted for 12.2% of all international development assistance (ODA) (USAID 2005a, 2008a). The expansion came with a renewed ideological vigour. Endowment reports and statements by Gershman throughout the 2000s, who continued to head the endowment, targeted Venezuela, Bolivia, and Ecuador as the main enemies of democracy (see Gershman 2000, 2008). At the first meeting of an anti-leftist regional affiliation of the NED's world movement for democracy in Panama in February 2008, the Latin American and Caribbean Network for Democracy, Gershman (2008) announced the ideological battle in no uncertain terms. Commenting on the spread of democracy in the 1990s, he cautioned about the backsliding that took place in Venezuela through 'the relentless concentration of executive power' and the 'insurrectionary toppling of constitutional governments' in Ecuador and Bolivia. 'To the extent that this assault on democracy took place under the cover of populist demagoguery and avoided blatant military coups and open forms of repression such as imprisonment, exile,

---

3   Under Reagan, Abrams was Assistant Secretary of State for Human Rights and Humanitarian Affairs in the early 1980s and later Assistant Secretary for Inter-American Affairs. He was found guilty for his involvement in the Iran-Contra Affair but later pardoned by President George H.W. Bush. John Negroponte was US Ambassador to Honduras from 1981 to 1985 when that country was being used as a staging ground to train the contras in Nicaragua and the death squads of El Salvador. Otto Reich was Assistant Administrator of the US Agency for International Development (USAID) in charge of US economic assistance to Latin America and the Caribbean in the early 1980s. Reich also established and managed the inter-agency Office of Public Diplomacy for Latin America and the Caribbean between 1983 and 1986, which disseminated propaganda against the Sandinistas (Grandin 2006).

and torture,' Gershman warned, 'many tended to ignore the backsliding or even to develop an attitude of tolerance towards it'.

Against this backdrop, democracy assistance remained an important vehicle to exert political influence through hard tactics intended to build civil society opposition forces to wage the war of position. In Venezuela, critics denounced the links between US democracy programmes and opposition groups in the early 2000s. When one endowment-funded NGO, *Súmate*, launched a recall referendum campaign against Chávez in 2003, the government struck back by charging it with treason and conspiracy (Allard and Golinger 2009) (the move was criticized, but one could only imagine what would have happened if Chávez began funding opposition groups in the United States to counter the democratic backsliding of the Bush administration). Chávez had good reason to be sceptical of US intentions; in 2002, when forces within the military aligned with the traditional oligarchy temporarily seized the reins of power, the United States immediately recognized the new government – in blatant violation of the OAS Democratic Charter – before the coup was swiftly reversed. *The Guardian* revealed that, in the lead up to the coup, Reich had met with the plotters at the White House to discuss the logistics of the putsch (Emerson 2010).[4]

In Haiti, the IRI's role in organizing the opposition to the government of Jean-Bertrand Aristide was so controversial that even the US ambassador – a holdover from the Clinton years – publicly denounced the group (discussed in Chapter 3) (Bogdanich and Nordberg 2006). He was ignored by his boss at the State Department, Otto Reich, which used IRI to advance a covert policy of regime change that eventually succeeded. In Bolivia, President Evo Morales denounced USAID for supporting opposition groups on numerous occasions. Declassified government documents obtained by one investigative reporter substantiated claims that democracy promotion programmes were explicitly formulated to undermine Morale's party, the *Movimiento al Socialismo* (Movement towards Socialism – MAS) (discussed in Chapter 5) (Bigwood 2006). In September 2008, Morales expelled US Ambassador Philip Goldberg from the country for supporting right-wing autonomist forces in the eastern lowlands. In Nicaragua, during the 2007 election campaign, US Ambassador Paul Trivelli threatened Nicaragua with economic sanctions if Sandinista leader, Daniel Ortega, were elected president (Emerson 2010).

The reaction to the new interventionism in Latin America occurred within the framework of a larger global rejection of US democracy promotion. Liberal proponents of democracy promotion such as Thomas Carothers (2006) have attributed this backlash to the unpopular war in Iraq as well as the blowback of autocrats, particularly in the former Soviet bloc, where US democracy promotion was linked to the so-called colour revolutions in Georgia, Ukraine, and Kyrgyzstan.

---

4   IRI issued a press release in support of the coup while its Democratic counterpart, the NDI, remained silent. To its credit, the NED expressed its disagreement with IRI for voicing its support of an unconstitutional action (Weisbrot 2009).

While this is partially true, the narrative overlooks legitimate concerns raised by how democracy assistance is being used to accomplish regime change in countries that are not undemocratic or at least no less democratic than regional allies. By abstracting democracy promotion from foreign policy interests, it obfuscates how the backlash is linked to the larger nationalist resurgence and its rejection of the neoliberal Washington consensus.

Although those movements and governments contesting neoliberal-polyarchy have been dismissed as populist and 'anti-systemic', US policymakers have in part registered the backlash as the consequence of increasing global inequality. A report commissioned by the Senate Foreign Relations Committee Report, for instance, warned that the inability of Latin American governments to 'adequately convert constituents concerns into responsive laws and policies is one important factor driving the poor and the politically marginalized toward leaders who promise popular but often short sighted solutions' (United States Senate 2006: 7). Speaking on democracy promotion in Latin America before the house international relations committee in 2006, Paula J. Dobriansky, under-secretary for democracy and global affairs under both the Bush and Obama administrations, summed up the ongoing conventional wisdom: 'We will work with all governments from the left, from the right, as long as they are committed in principle and practice to the core conditions of democracy, to govern justly, to advance economic freedom and to invest in their people.' This greater concern with economic inclusion was reflected in a new democracy and governance framework for USAID (2005a), which emphasized the importance of social rights and making democracy 'work' for marginalized groups such as women, youth, and minorities. The new language informing USAID's approach to democracy promotion reflected the emerging post-Washington consensus, which acknowledges the need to tackle poverty and inequality as objectives in their own right (Onis and Senses 2005).

Yet, with the failure of the FTAA talks (1998–2004), the Bush administration sought to strengthen the neoliberal regime of accumulation through the expansion of free trade agreements (FTAs) to promote trade and investment which lock in reforms beneficial to US capital. Agreements were negotiated with Chile (2004), Central America (Costa Rica, El Salvador, Guatemala, Honduras, and Nicaragua) and the Dominican Republic through the DR-CAFTA, and Peru (2009).

USAID Administrator for Latin America and the Caribbean, Paul J. Bonicelli (2007), acknowledged the strategic reasons behind the new concern with issues of social justice at a speech at conference organized by the Center for International Private Enterprise in Lima, Peru, when he proclaimed, 'frustration with the failure of democratic institutions to deliver improved standards of living may further enable the purveyors of populism'. The location of Bonicelli's speech was symbolic, as the government of Alejandro Toledo pioneered many of the institutions and practices that defined the inclusive neoliberal approach which came to define US soft tactics (discussed in Chapter 4).

The new hegemonic discourse was underpinned materially by massive amounts of democracy assistance to allied states. Although these escaped critical attention, they were central to both the Bush administration's attempts at stabilizing conservative governments seeking to prevent structural change and, as we shall see, remain just as important under the Obama administration. USAID democracy assistance flows during the Bush years as illustrated in Figure 2.1 provide some indication of the importance of stabilizing allies – during Bush's first term, Colombia was by far the largest recipient followed by Peru, Mexico, and Ecuador (at the time still an ally). Conversely, democracy assistance to destabilize enemies was much more modest; Cuba was the fifth most important destination, but no actual programming occurred on Cuban soil.[5]

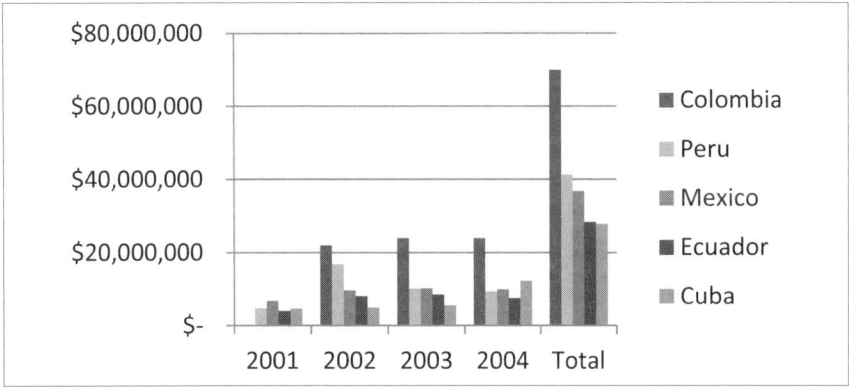

Figure 2.1 Top five destinations of USAID democracy assistance during George W. Bush's first term

During Bush's second term, Colombia, Mexico and Cuba remained in the top five, but Peru and Ecuador were replaced by Bolivia and Haiti, as illustrated in Figure 2.2. Bolivia's increased prominence coincided with a new strategy of regime change; Haiti, however, only became a major destination, after the coup had occurred – in 2004, $3.6 million was spent on democracy assistance, an amount dwarfed by the $37.8 million allocated the following year.

---

5   Organizations which receive USAID funds for democracy promotion in Cuba are largely US-based. A 2006 report by the Government Accountability Office offered stinging criticism of USAID for its lack of oversight of the programme. The report noted that nearly all of the $74 million spent on contracts through the programme, which was established by the Clinton administration in the mid-1990s, had been distributed without competitive bidding or oversight in 'a programme that opened the door to waste and fraud'. A *Miami Herald* article the same year noted that most of the USAID money remained in Miami or Washington, creating 'an anti-Castro economy that finances a broad array of activities' (cited in Collins 2010).

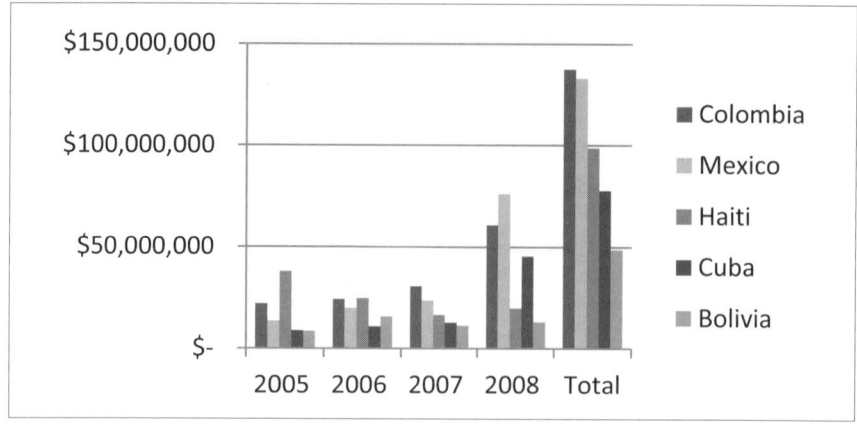

Figure 2.2 Top five destinations of USAID democracy assistance during George W. Bush's second term

Although data on NED grants are not available for Bush's first term, the figures from 2005–2010 illustrated in Figure 2.3 reveal that the QUANGO followed a similar pattern of focusing on both key US allies and regional enemies. The top five destinations for grants during this period were Cuba, Venezuela, Colombia, Mexico and Peru.

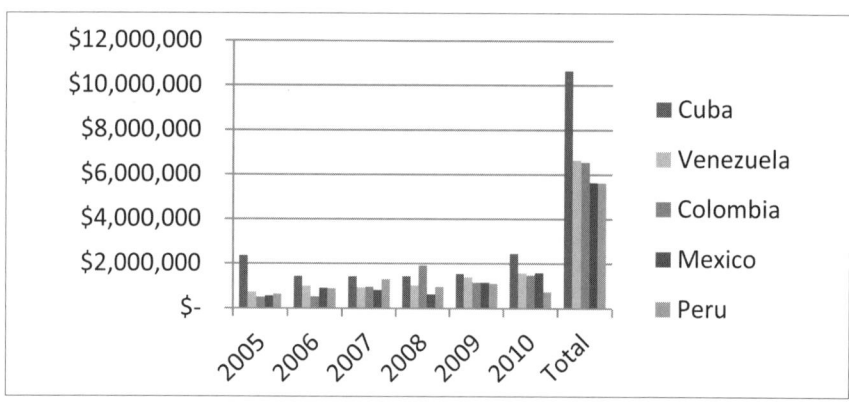

Figure 2.3 Top five destinations of grants from the National Endowment for Democracy (2005–2010)

But even where democracy promotion has been linked mainly to soft tactics, securitization has been its coercive underbelly. Indeed, the emphasis on soft tactics and new hegemonic discourses coincided with the rise of the Pentagon in regional policymaking within the optic of the War on Terror. Military and police aid increased, and more Latin American soldiers were trained in the United States between 2001 and 2005 than in the previous 50 years.[6] The Pentagon's 'effective sovereignty' policy contended that US security was threatened by the failure of states to exercise control over ungoverned spaces within their borders. Terrorists, narco-traffickers, arms traffickers and document forgers were identified as the new enemies, and military aid continued to fund counter-insurgency activities in Colombia and other regional allies (Emerson 2010).

Security issues also took centre stage amongst the NAFTA partners and came to define North American deep integration. In March 2005, the three countries met in Waco, Texas and agreed to form the Security and Prosperity Partnership (SPP) within the NAFTA framework. The agreement built on the logic of Bush's *National Security Doctrine* with its insistence that free markets, economic growth, prosperity and national security are positively correlated. The North American Competitiveness Council was formed within the SPP of executives of the region's largest and most powerful transnational corporations to represent the interests of business in the working groups; there is no representation of labour, environmental or citizen's groups (Carlsen 2009a).

The reaction to the interventionism of the Bush administration also led many liberals in the democracy assistance field of practice to call for a more multilateral approach. This view was also reflected in a Senate Foreign Relations Committee report, which recommended that USAID and the endowment strengthen partnerships with Latin American, European, and international organizations in implementing democracy promotion programmes in the Americas. The presidents of IRI, Lorne Craner, and NDI, Kenneth Wollack, argued that the concept of democracy promotion must be rejuvenated by working 'with allies and through international organizations to give such efforts greater legitimacy and an international face' (Craner and Wollack 2009). Speaking specifically about Canada, Gerald Hyman, the former director of USAID's office of democracy and governance, noted that Canada can be active in many places the US cannot, such as Cuba (Bird and Drolet 2009). Such statements indicate that at least some policymakers in the United States are increasingly looking to US allies to give a more universalistic quality to declining *Pax Americana* through their own democracy promotion efforts. The appeal has not fallen on deaf ears: with the growing stake of Canadian multinationals in the global economy, Ottawa has become increasingly inclined to use its own democracy promotion efforts to defend an unequal regional order of which both countries are prime beneficiaries.

---

6   Between 2001 and 2005, 85,820 Latin American soldiers were trained in the United States compared to 61,000 soldiers and police trained by the infamous School of the Americas from 1946 to 2000 (Tokatlian, cited in Emerson 2010).

The Liberals Define a more Muscular Middlepowermanship

With the collapse of the twin towers, Canada followed the US lead in adopting a more instrumentalist approach to democracy promotion based on its security objectives and support for free markets under the Liberal governments of Jean Chrétien and Paul Martin. This coincided with an increased in both CIDA's democracy assistance budget, which doubled over 10 years from $223 million in 1996 to $477.9 million in 2006 by the time the Conservatives came to power (Government of Canada 2007).

The Liberal's International Policy Statement relegated the Americas to a role of minor importance, though Canada continued to partner with a range of countries in the region, particularly Haiti, which began to receive the majority of ODA. Even before the devastating earthquake in early 2010, Haiti was receiving more than 40% of Canadian aid in the Americas, ranking it well above second-tier countries like Colombia, Peru, Bolivia, and Honduras (Randall 2010b). The neoliberal agenda was also pursued through an FTA modelled after the NAFTA and its Chapter 11 provisions surrounding investment with Costa Rica (2002) and negotiations with four Central American countries and the Caribbean Community which remained stalled (see Shamsie 2007).[7] The Liberals also prioritized a small number of countries for democracy assistance which remained the same throughout the decade. As illustrated in Figure 2.4, these were Haiti, Bolivia, Peru, Jamaica and Guatemala. Again, Haiti has received considerably more than the other recipients – from 2001 to 2009, total democracy assistance equalled $139.3 million; Bolivia, the second largest recipient, ranked far-behind at $43.5 million.

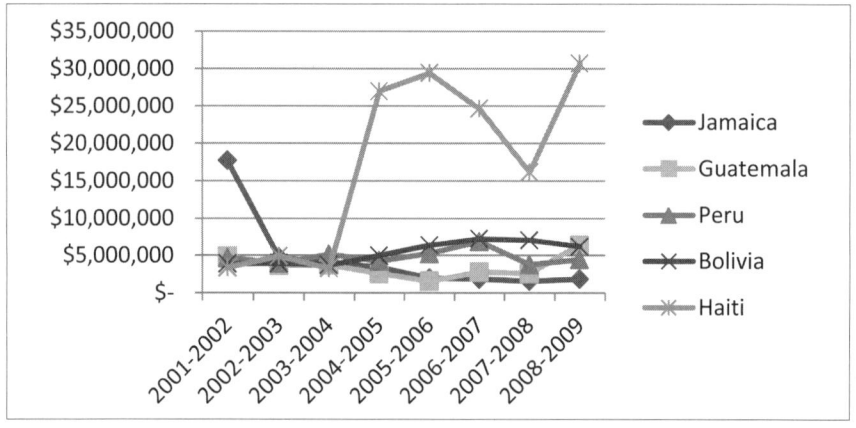

Figure 2.4 Top five destinations of CIDA democracy assistance from 2001–2009

---

7 An FTA with Chile was signed in 1997.

Through initiatives such as the 'Responsibility to Protect' doctrine – formulated initially by the UN-sponsored international commission on intervention and state sovereignty led by Canada under the Chrétien government – Canada began developing its own brand of middle-power humanitarian interventionism as an extension of its concern with human security. Under the Martin government, Canada's 'moral imperative' to act to prevent abuses such as ethnic cleansing and genocide was further developed to include failed and fragile states, which were deemed threats to the security of their citizens and the international community. The doctrine was accompanied by a new whole-of-government approach to providing humanitarian and reconstruction assistance through increased coordination between Foreign Affairs, the Department of National Defence, and CIDA. To coordinate whole-of-government policy and programme engagements in fragile states – including support to democracy – the Stabilization and Reconstruction Task Force (START) was established within Foreign Affairs.

Canada's concern with preventing humanitarian disasters from occurring in fragile states could be read as an extension of traditional Pearsonian internationalism. Critics such as Thede (2008), however, have argued that the responsibility to protect defines a 'hard security' approach which can be used on ideological grounds in countries where threats may in fact be minimal. This has occurred while the concept of human security advanced by previous Liberal governments has been progressively narrowed to prioritize physical security over socio-economic and development considerations. Instability itself, moreover, is often indicative of the failure of the neoliberal development paradigm.

This was certainly the case in Haiti, where the imposition of neoliberal restructuring by international actors – a process supported by Canada – played an important role in undermining the state and creating the conditions for the crisis of the early 2000s. There, Canada invoked the responsibility to protect only after the democratically-elected government of Jean-Bertrand Aristide was dislodged by paramilitary forces to justify a UN mission that stabilized the situation in cooperation with a repressive interim government. The doctrine became a means of legitimizing a strategy of regime change to which Canada had also contributed through various democracy promotion initiatives.

The liberal interventionist doctrines also help reinforce the hegemony of regional and world order. As Neufeld (2004) argues, Canadian middlepowermanship provides *Pax Americana* with a more universalistic quality while advancing the mutual interests of multinational corporations of both countries which benefit from a more stable (and liberalized) world order. While material interests were undoubtedly central to the Liberal embrace of the new interventionism, the search for international prestige also played a role, particularly in Haiti (as discussed in Chapter 3). Yet there is also a cultural-institutional dimension to Canada's embrace of a more aggressive middlepowermanship. As Charbonneau and Cox (2008) argue, decades of Canadian-US defence integration have produced a 'common sense' understanding of what constitutes 'good' and 'evil' within a self-help anarchic world of state actors. Canadian military culture 'is an integrated facet of American militarism', a factor that helps explain Canada's decision to follow the United States into Afghanistan, where its own interests are minimal.

Below these macro-level trends redefining Canada's place in the world, the Liberal governments led a more concerted effort to centralize the democracy assistance field of practice. Major initiatives included the establishment of the 'democracy council' bringing together the main organizations that deliver Canadian democracy support under the auspices of CIDA and Foreign Affairs and bureaucratic restructuring within both departments. In the early 2000s, discussions also began around the idea of creating an umbrella foundation to provide grants to Canadian political parties to support democratic development abroad. The proposal was adopted by the parliamentary Special Committee for Foreign Affairs and International Development, which conducted a review of Canada's role in international support for democratic development (House of Commons 2007). The report of the committee further emphasized the traditional rights-based orientation but failed to acknowledge how Canadian foreign policy had not lived up to this standard. It also missed the opportunity to reflect on the significance of the backlash against US democracy promotion, which it followed the US lead in attributing to self-interested autocrats and the challenges in Iraq. Yet, despite the growing contradictions of Canada's approach to democracy promotion, which in large part reflected the ongoing shift from humane to neoliberal internationalism, the Liberal government did not directly attack the autonomy of more progressive NGOs delivering democracy assistance. It would fall to the Conservative government of Stephen Harper to deal this final blow.

Strengthening the Regional Right: Conservatives and Democrats

Obama–Harper Convergence

Despite the shifts in the fortunes of political parties in both Ottawa and Washington, continuity in the area of foreign policy marked the transition from Liberal to Conservative and from Republican to Democratic governments in the mid-2000s. Whereas Obama may have tempered the rhetorical bluster of Bush's neoconservatives, military escalation in Afghanistan and drone attacks in Pakistan marked the ongoing geopolitical approach. In the president's March 2011 address – which laid out the so-called 'Obama Doctrine' – the administration elaborated a new rationale for military intervention in Libya based not on ensuring its safety but on defending its 'interests and values' (Diamond 2011). And while Harper seemed to turn his back on Liberal multilateralism, his policies are the logical continuation of the primacy of neoliberal expansion that drove Canadian foreign policy away from the enlightened self-interest of the Trudeau years. Both states have retained the right to promote democracy by military means, even though the outcomes on this front have been catastrophic.[8] What has changed, however, is the ability of either country to intervene in the affairs of the Americas, where the power dynamics have shifted radically.

The Harper government led the way in responding to the new regional sensibility through a policy of re-engagement designed, in part, to enhance Canada's international status.[9] Harper – a hard-line conservative once committed to dismantling the Canadian welfare state – has sought to position Canada as a more ideologically moderate northern power. In an interview with *Americas Quarterly*, Harper noted that 'there is a perception in the region that there exist only two models for development, one that is focused on social justice, the other on economic liberalization. Canada offers a different model, and illustrates that these are not mutually exclusive' (quoted in Cameron and Hecht 2008).

The centrepiece of the government's approach has been the Americas Strategy, which prioritizes democracy, prosperity, and security (democracy was also identified by the government as one of four core values guiding foreign policy, along with freedom, human rights, and the rule of law). It adopts the rhetoric of both Bush's National Security Doctrine and the SPP (in terms of the alleged positive relationship between the three pillars) as well as the secrecy of the latter (the actual strategy has never been made public).[10] Yet the concern with democracy seems to be largely rhetorical. In addition to its alliances with the gravest human rights abusers in the region (discussed below), the Harper government has opposed legislation that would subject Canadian mining companies to binding regulations,[11]

---

8 Neither the governments of Barrack Obama and Stephen Harper, moreover, can be considered models of democracy, even by their own liberal-democratic standards. The Obama administration has done little to repeal the assault on liberal-democratic rights contained in Bush-era legislation such as the Patriot Act, for example, and the first two minority governments under Prime Minister Harper were well known for their antipathy towards parliament and the independence of government entities and agents of parliament (see Martin 2010).

9 According to a wikileak cable, Harper was inspired by Australia's engagement with its neighbouring countries, where it came to play an important regional role despite its lesser status in global affairs (Canadian Press 2011).

10 An online summary of the initiative on the DFAIT website makes reference to an 'Americas Strategy Implementation Plan'. An Access to Information Request submitted for a copy of the strategy and the implementation plan was denied based on article 69(1), which states that the act does not apply to confidences of the Queen's Privy Council for Canada (i.e., Cabinet). The exclusion from public consumption of any documentation on one of the most important foreign policy strategies to be put forward by Canada in recent years is disconcerting, especially by a government that preaches transparency and claims to be promoting democracy.

11 The legislation in question was Bill C-300, which codified a number of recommendations made by an advisory group on Corporate Social Responsibility and Canadian mining which included representatives of the industry and civil society. Among other things, the legislation would have regulated several government agencies that support

and has continued to push a vision of prosperity rejected by social movements and civil society actors across the region. As Canadian mining companies are increasingly criticized for their actions throughout the region (Gordon 2010), the notion that Canada offers an alternative model that can appeal to other countries reflects the wishful thinking of a policy elite increasingly out of touch with reality.

Speaking before the parliamentary Special Committee for Foreign Affairs and International Development on Canada's role in democracy promotion, then-Minister of Foreign Affairs Peter MacKay stated that 'Canada's reputation as a fair player confers clear advantages: we were never a colonial power; we do not have great power ambitions; our motives are not suspect; our agenda is not hidden'. Perhaps its agenda is no longer hidden, but Canada is increasingly behaving like a colonial power.

The Harper government has also begun playing a junior role in the US War on Drugs in the Caribbean. Since 2006, Canadian Forces have participated in numerous counter-narcotics missions in the Caribbean as part of the wider US Joint Interagency Task Force-South through Operation Caribbe. Between October and November 2010, the HMCS St. John's Canadian naval ship intercepted a self-propelled semi-submersible submarine filled with 6,700 kg of cocaine, with a street value of $180 million (Edmonds 2012, National Defence 2011). The Canadian government is also providing financing for the training of elite forces through Caribbean Junior Command and Staff College at military facilities in Jamaica (Edmonds 2012). Harper's Anti-Crime Capacity Building Program (ACCBP), launched in 2009, is also providing up to $15 million a year to enhance the capacity of government agencies, international organizations and non-governmental entities throughout the Americas. In Central America and the Caribbean, the ACCBP is focusing on preventing illicit drug trafficking and reforming the security sector (DFAIT 2011c).

The approach of the Obama administration has followed along similar lines. In terms of his policy team, Obama retained Tom Shannon, Bush's Assistant Secretary of State for Western Hemispheric Affairs, before nominating him as Ambassador to Brazil. He was replaced by Arturo Valenzuela, who was one of the authors of Plan Colombia when he worked in President Clinton's administration. The new administration tempered the rhetoric against the regional left and initially relaxed provisions in the embargo against Cuba which some hoped signalled a meaningful policy shift (for example, Erikson 2011). Secretary of State Clinton (2010) also re-situated democracy promotion efforts through a new emphasis on the mutually reinforcing relationship between 'development, democracy and human rights' rather than a blind faith in trickle-down economics. Yet, as Vanderbush (2011) points out, Obama's 'good neighbour imperialism' has largely remained faithful to Republican military, commercial and geo-political objectives.

---

extractive companies, including Export Development Canada, the Canadian Pension Plan, Canadian Embassies and trade commissioner staff. Companies seeking support from these entities would have had to have conformed to a set of binding standards (Keenan 2010).

Indeed, both Canada and the United States have prioritized bilateral relations with right-wing governments in Colombia, Peru and Mexico, which have used heavy doses of repression to deal with political unrest, crime and social problems (Peru will be discussed in Chapter 4). Conversely, they have selectively condemned ideological enemies for being undemocratic (they have also cemented ties with 'responsible' left governments such as Chile while courting Brazil to develop closer economic ties). Former Minister of State for Latin America, Peter Kent, was particularly outspoken on following a trip to Venezuela in January 2010, in which he failed to meet with any representative of the government. Kent declared: 'there's no question the democratic space in Venezuela is shrinking' (Berthiaume 2010a). On his weekly TV programme, *Alo Presidente*, Chávez retorted that he would not take lessons from an 'ultra-right' government that 'closed' parliament, referring to the Harper government's prorogation of the House of Commons (Gordon and Webber 2010). Meanwhile, DFAIT (2011b) has claimed that 'many countries in the Americas like Brazil, Chile, Colombia and Peru have taken the road to democratic governance and consolidation of economic gains based on open markets and enlightened macro-economic policies'.

The Obama administration toned downed hostilities with Venezuela, but tensions remained. After diplomatic relations were restored between the United States and Venezuela, the conflict resurfaced when Obama's nomination for US ambassador, Larry Palmer, was rejected by Chávez for commenting that Venezuelan military's morale was low and that the FARC may be finding refuge in Venezuela. Obama has refrained from directly attacking Chávez, but US democracy promotion agencies regularly lambaste the leader.[12] Freedom House (2009) has grouped Chávez together with '21st Century Authoritarians' in China, Iran, Russia and Pakistan, which it condemns for employing soft-power methods to advance their interests internationally, thereby 'tilting the scales toward less accountable and more corrupt governance across a wide swath of the developing world'. IRI (2008a) claims that: 'in Latin America, where democracy has been threatened in countries like Venezuela and Bolivia, Colombia remains dedicated to building democratic institutions, making it a critical player in the hemisphere's geopolitical landscape'.

Policies pursued by the government of President Álvaro Uribe in Colombia, however, can hardly be considered democratic. Uribe's security policy, for instance, led to widespread human rights abuses such as extrajudicial killings and forced disappearances by the military (Rojas 2009). The most egregious example included the 'false positive' scandal in which soldiers were found to be killing civilians and dressing them in rebel uniforms or giving them guns in order to pass them off as guerrillas or paramilitaries. A perverse incentive structure rewarding

---

12   Obama has resorted to indirect attacks, however, such as when he spoke of 'some leaders who cling to bankrupt ideologies to justify their own power' while praising the overall progress of democracy in the region at a speech in Santiago in March 2011 (Perez-Rocha 2011). Chávez has typically been more direct in his rebukes of the president.

officers for body counts provided the institutional context in which this abuse took root (McDermott 2009). And yet, Colombia remains the largest recipient of military aid in the region (see below). The US–Colombia Defense Cooperation Agreement, which came into force in October 2009, facilitates US access to several Colombian military facilities with the aim of coordinating actions in the War on Terror and counter-narcotic operations. Soon after the provisional agreement was signed, several Latin American presidents protested at a meeting of the newly-established UNASUR. Ecuadorean president Rafael Correa (whose country was bombed the year before by Colombia in an extraterritorial operation against the FARC) provided the most powerful voice of opposition, insisting on the need to coordinate anti-narcotic operations without the assistance of the DEA or the US military (Phillips 2009). After closing a US military air base in Mantra, Correa noted: 'We can negotiate with the US about a base in Mantra if they let us put a military base in Miami' (Robinson 2011). Obama also reactivated the US Fourth Fleet in the Americas after 58 years of absence, providing further indication that the United States is reaffirming its military preponderance in Latin America (Carranza 2010).

Both Canada and the United States recently signed FTAs with Colombia despite the country's horrendous labour record (Colombia is the most dangerous place in the world for trade union activists; in 2010, 51 were reportedly killed).[13] For the United States, the agreement was particularly controversial given Obama's official rejection of a free trade deal with Colombia during his presidential campaign, which he denounced for the routine assassination of union activists. Indeed, Colombian labour unions activists have consistently opposed the proposed agreement, including before the US congress.

The militarization of the drug war in Mexico by President Felipe Calderón – massively funded by the United States under the Merida initiative (also known as Plan Mexico for its similarities with Plan Colombia) – has also led to widespread abuses. The initiative, originally proposed by Bush in the framework of the North American SPP, coincided with the deployment of 45,000 soldiers throughout the country. Mexico's National Human Rights Commission reported a six-fold increase in human rights abuses committed by soldiers during the first two years of the Calderón administration, which surged from 182 in 2006 to 1,230 in 2008 (Carlsen 2009a, Mercille 2011). Between 2008 and 2010, the United States provided Mexico with $1.3 billion in security-related assistance (Witness for Peace 2011). The plan, which does nothing to deal with the root causes of drug production and trafficking – namely poverty and insatiable demand in the United States – has been strongly supported by Obama, who led efforts to obtain congressional renewal

---

13   It bears mention that Obama's Attorney General, Eric Holder, was counsel to Chiquita bananas at a time when it was allegedly involved in ordering hits against labour militants by Colombian paramilitaries (Cray 2010). It should be noted that, in Guatemala, murders of unionist increased following passage of CAFTA as workers' rights were seen as economic liabilities in an environment of heightened competition (Carlsen 2011).

beyond the plan's expiration date (the State Department also claimed that Mexico was complying with human rights conditions over the protests of civil society organizations and members of congress in both countries). 90% of the guns that fuel the drug wars also come from the United States, providing a lucrative economy for arms companies who profit from both the trade and the militarized strategy of suppressing it.[14] Although the motives of the administration are unclear, the War on Drugs is likely being fuelled by domestic anti-drug sentiment and the regional military presence that the policy affords. As a result of the disastrous consequences of this strategy, however, a growing number of Latin American leaders are calling for discussions on legalization, including former presidents of Brazil, Mexico and Colombia, as well as the current presidents of Colombia and Guatemala. Although the Obama administration has claimed that it is open to dialogue, Vice President Joe Biden recently insisted during a visit to the region that legalization is not an option (Carlsen 2012).

Even more worryingly, the United States and Canada helped legitimize the coup against President Manuel Zelaya in Honduras by recognizing the results of fraudulent elections hosted by the de facto government of Roberto Micheletti in November 2009, which were in violation of an internationally-brokered compromise and opposed by the rest of the hemisphere (Carlsen 2009b). Secretary of State Clinton pushed for the reintegration of Honduras into the OAS, citing supposedly free and fair election and the establishment of a truth commission. Minister of State Peter Kent took the same position (see Gordon and Webber 2011). At the 38th Mercosur Presidential Summit held in Montevideo on 8 December 2009, the five Mercosur partners strongly condemned the American recognition of the elections (Carranza 2010). Zelaya was overthrown by the military (with the support of Congress and the Supreme Court) for pushing a constitutional referendum, which his critics alleged was intended to allow a second presidential term.[15] Zelaya, himself a wealthy rancher, had alienated the traditional oligarchy through modest reforms and by joining Chávez's ALBA. In the case of the United States, democracy promotion and the old connection between the American political elite and the Honduran right most certainly

---

14   In a scandal that recently came to light, the US Bureau of Alcohol, Tobacco, Firearms and Explosives was found to have allowed illegal buyers to smuggle weapons into Mexico through operation 'Fast and Furious'. The point of the operation was to track the guns back to cartel leaders who were believed to be the ultimate buyers. After a US agent was killed with one of the guns, Attorney General Eric Holder insisted that he had not known about the operation; soon after, President Obama expressed his full confidence in the attorney. However, the Republican Party obtained a series of five reports to Holder from Michael F. Walther, director of the National Drug Intelligence Center, making reference to the operation (Serrano 2011).

15   In fact, the non-binding poll would have asked Hondurans whether they were in favour of an additional ballot to be added to the November 2009 elections, which would have given them the opportunity to vote for or against convoking a National Constituent Assembly to draft a new constitution.

played a role in laying the groundwork for the coup and in portraying it as a defence – rather than a violation – of the rule of law (see Golinger 2009).[16]

The election led to the victory of Porfiro 'Pepe' Lobo – defeated by Zelaya in the 2005 election – who quickly set about reinstating the neoliberal agenda behind the coup and who has presided over a harsh climate of repression (Joyce 2010). Following a visit to Honduras in May 2010, the Inter-American Commission on Human Rights expressed serious concerns about human rights abuses and the impunity with which they were being committed. The *Comité de Familiares de Detenidos Desaparecidos en Honduras* (Committee of Relatives of the Disappeared in Honduras – COFADEH) has reported multiple politically-motivated killings under the Lobo government, which recently resorted to extreme repression to deal with striking schoolteachers (Frank 2011). Canada recently completed FTA negotiations with the Lobo government despite its poor human rights record. Canadian officials have also begun advising the Honduran government on a new mining law based on the non-binding principles of a CSR framework that will undoubtedly benefit Canadian mining companies. In May 2009, the Zelaya administration had completed a draft mining bill which would have imposed tax increases, prohibited open-pit mining and the use of toxic substances such as cyanide and mercury, and required prior community approval before mining concessions could be granted. The law was developed in part to respond to environmental and social concerns surrounding the mining activities of Canadian-based Goldcorp's San Martín open pit gold mine. The coup put an end to this initiative, along with what DFAIT has characterized as the 'anti-mining Zelaya administration' (Moore 2012) (Canadian interests are also prominent in the assembly-manufacturing sector, with Montreal-based Gildan Activewear running several sweatshops in *maquiladora* zones in the country).

Canada and the United States have not always acted in lockstep, however, and the Harper government has maintained positive relations with Bolivia, which stands in sharp contrast to the acrimonious bilateral relationship between Bolivia and with the United States. Canadian mining interests are growing in Bolivia, and it is quite likely that Canada has chosen a more cautionary approach (even though, as we shall see in Chapter 5, it sought to stabilize the neoliberal government that came before Morales' sweep to power). But the tendency, on the whole, has been for an increasingly convergent approach that reflects a distinct North American vision of how things ought to be.

---

16   Key neo-conservatives who lobbied on behalf of the coup government include Roger Noriega, Otto Reich and Daniel W. Fisk, who, among other high-ranking positions, was the Deputy Assistant Secretary of State in the Bureau of Western Hemisphere affairs under President George W. Bush (Thompson and Nixon 2009). Mr. Fisk's defence of the de facto government is particularly noteworthy given his role as Coordinator for Governance for IRI, a position he assumed in September 2009.

Democracy Assistance: Stabilization and Retrenchment

Shifting approaches to democracy assistance under the Harper government and the Obama administration have paralleled these larger forms of hemispheric engagement. Although destabilization of ideological enemies is increasingly unfeasible in the current historic conjuncture, democracy assistance is being channelled towards allied states which have prioritized the neoliberal-security agenda. Canada and the United States, in short, are increasingly adopting a model of stabilization as the new *modus operandi* of democracy assistance. For Canada, this has coincided with a restructuring of the field of practice and the re-articulation of the conception of democracy that guides its engagement.

Under the Obama administration, overall democracy assistance through USAID in the Americas reached over $600 million in 2010, or less than half the over $1.3 billion spent on peace and security assistance in the same year. The main recipients are Mexico, Colombia, and Haiti. In 2010, they respectively received $209 million, $50 million, and $186 million (USAID 2011a). In each of these countries, there is a strong focus on supporting the state, though there are also civil society and political-party strengthening programmes. Venezuela continues to be a comparatively marginal recipient of democracy assistance through USAID (but not the NED), and in Bolivia democracy programmes have been significantly curtailed (in 2010, funding actually came to halt in Bolivia as the conflict with the Morales government escalated; in Venezuela, $6 million was spent).

The tendency, in other words, is to support security-state polyarchies function more effectively. Unable to secure even a decadent form of hegemony – which would require some degree of class compromise - the main recipients of democracy assistance have increasingly relied on transnational support to govern through a strategy of passive revolution. Yet the new policy approach is defined more by coercion than by consent – spending on security considerably outweighs democracy assistance. In Mexico, security and peace assistance was just under $530 million in 2010; in Colombia, it was $516 million in the same year (USAID 2011a). Stabilization is deeply related to pacification, and the recourse to a critical focus on soft tactics should not mask the violence that underpins US geopolitical power in the Americas.

As for Canada, Latin America as a whole has increased in importance as a source of ODA in the context of CIDA's list of 20 'core countries' which dropped many poor African countries in 2009 and added wealthier ones in Latin America and the Caribbean, notably those of particular trade interest to Canada (soon after the list was released, CIDA President Margaret Biggs listed for the first time Canada's foreign policy considerations as an explicit official criterion for selecting core recipients) (Brown 2011). Colombia, which received a total of $15 million in ODA in 2010, is now the second largest recipient of Canadian ODA after Haiti, which was allocated $100 million the same year (Randall 2010b). Haiti, Colombia and Guatemala also comprise the three priority countries of DFAIT's

Strengthening Security, Rule of Law and Democracy in the Americas programme, managed by START.[17]

Harper and the Re-Organization of the Canadian Field of Practice

These broader trends in development have intersected with important changes in the bureaucracy, with DFAIT increasingly displacing CIDA as the central actor in the Canadian field of practice. In mid-2006, Foreign Affairs created a small democracy unit, as well as a democracy and governance division. The department also established a democratic transitions fund under the Glyn Berry programme designed to deal with violence, political instability, and democratic backsliding, particularly in the Andean region. At the time of writing, the Andean Unit for Democratic Governance is currently staffing positions in Peru and four neighbouring countries. The Unit will focus on 'providing support to democracy in the Region by developing networking contacts in Colombia, Venezuela, Ecuador, Bolivia and Peru'. It will be similar to a security hub being established in Panama (the Regional Office for Peace and Security), which will provide 'analysis and engagement' in support of combating drugs, crime and terrorism. Geographically, the focus will be Central America and the Caribbean (DFAIT 2011d).

Despite the rights-based language that describes the Glyn Berry Program (DFAIT 2008), the conceptual scope of democracy has been considerably narrowed in both theory and practice. Unlike CIDA's earlier Human Rights, Democratization, and Good Governance policy, the programme seems to ignore the importance of social and economic rights. One Foreign Affairs official who was interviewed stated that the department was increasingly focusing on representative democracy as defined by the OAS Inter-American Democratic Charter. She noted that even the emphasis on 'citizens' in the new programme had been controversial since it could be used to defend notions of 'popular democracy' advanced by 'populist regimes' such as the one in Venezuela. Canada has even provided support to US ideologically driven initiatives such as the previously mentioned regional affiliate of the NED's Movement for Democracy, the Latin American and Caribbean Network for Democracy.

Liberal critics argue that the Harper government is effectively dismantling Canada's tradition of strengthening democracy abroad, pointing to the many institutional changes that have occurred in the last six years.[18] Additionally, the

---

17   Personal correspondence with DFAIT official from the Glyn Berry Program (4 January 2012).

18   For example, DFAIT's democracy unit was recently absorbed by the Commonwealth and Francophonie division. The government also downgraded the importance of democracy promotion as a separate activity at CIDA. After re-branding the Canada Corps unit the office of democratic governance in October 2006, the activities of the office were integrated into CIDA's branches in 2009–2010. The Democracy Council, which grouped together the main actors in the Canadian field of practice to share information, has become defunct.

proposal for a political-party foundation, which the Conservative government initially endorsed by establishing an independent panel tasked with developing a model for the agency, has effectively been mothballed. While some critics warned that the agency would be used as a foundation for 'political warfare', it seems more plausible that the Harper government ultimately rejected the idea of supporting a multiparty foundation that would draw upon the expertise of all Canadian political parties.[19] Nor is it clear whether democracy promotion will be sidelined based on other foreign policy priorities – in December 2012, the Minister of State for the Americas launched an 'open dialogue' with 100 experts from academia and civil society to discuss the five-year old Americas Strategy, and many observers have suggested that the government will be reinforcing the democracy pillar of the strategy (though the results of the consultation will not be released to the public) (Bedrossian 2012).

If key institutions are being dismantled, however, the field of practice as whole is being reconfigured to conform to the government's political agenda. This is ultimately what has been behind the funding cuts of many NGOs that deliver democracy assistance. Early victims were Alternatives, a Montreal-based NGO that has been critical of Canada's involvement in Afghanistan, and the faith-based human-rights organization KAIROS, which has been critical of both Israel's actions in the Palestinian territories, Canadian environmental policy and the practices of Canadian mining companies (resulting in a scandal for the government as its beleaguered minister for international cooperation, Bev Oda, found herself under fire for misleading parliament about the decision to cut KAIROS' funding). The Forum of Federations, an independent international organization that was initially sponsored by the Chrétien government, has also lost funding. According to the president of the Canadian Council for International Co-operation, Gerry Barr, such developments 'represent a problem that is becoming more worrisome. Our organizations are being told to stop criticizing certain positions of Canada, or else their funding will disappear' (Castonguay 2010).

But the most controversial attack occurred against Rights and Democracy. In 2009–2010, the agency was the scene of an unfolding political drama in which some board members accused the Harper government of replacing colleagues who had stepped down with neoconservative ideologues. Under the leadership of the organization's chair, Aurel Braun, the new board members alleged, among other things, that the organization was pro-Palestinian, citing two grants made to one Israeli and two Palestinian NGOs critical of the Israeli government's human rights record. In January 2010, President Rémy Beauregard died of a heart attack after a tense board meeting on the future direction of the organization. The tragedy was followed by the resignation of several senior managers and staff in protest over

---

19   It is interesting to note, however, that one of the main advocates of the agency has been highly critical of the governments in Bolivia and Venezuela, calling upon the Canadian government to implement political party support programmes in these countries to encourage greater pluralism (Legler 2006).

the board's actions. As one board member stated: 'this resembled an ideological crusade, a witch hunt based on a good guys and bad guys narrative. You need a narrative of people who are the champions of freedom, or Israel, or democracy and liberty or whatever it might be.'

In the wake of the crisis, the Conservative government appointed Gérard Latulippe as the new president of Rights and Democracy. A former Quebec Liberal provincial cabinet minister who ran for the Canadian Alliance in 2000 and recently served as the head of NDI in Haiti, Mr Latulippe's political views fit well with the board's new mission. During the public consultation on 'reasonable accommodation' of immigrants in Québec undertaken in 2007, for instance, Latulippe warned that the concentration of Muslim immigrants undermined the proper functioning of Quebec society (Canadian Council on American-Islamic Relations 2010). In appointing him president, the government dismissed concerns voiced by all three opposition parties (Cheadle 2010). In April 2012, however, Foreign Affairs Minister John Baird announced suddenly that Rights and Democracy would be shut down as a result of the government's austerity cuts. Baird rather disingenuously cited the many challenges the agency had faced, noting that the announcement 'gives us a clean slate to move forward (DFAIT 2012). Although the agency under Latulippe's leadership would likely have followed the Conservative script, the Harper government seems to have decided that DFAIT, which absorbed the agency's functions, should take on the pivotal role in Canadian democracy promotion efforts.[20]

In many ways, these attacks are an extension of the Prime Minister's Office's policy of tightening control over governmental departments, officers of parliament and the public service into civil society. They serve, in large part, to undermine the ability of NGOs and other agencies to engage autonomously in international development, including in the provision of democracy assistance. They are perhaps as much about ideology as they are about the politics of control and centralization. The recent funding cuts to the Foundation for Canada and the Americas (FOCAL), a research institute that has traditionally shared the government's preference for free markets, suggest as much.

But they are also about replacing the traditional development field of practice with more compliant networks outside of the state. Indeed, the funding cuts have coincided with funding announcements for NGOs willing to team up with Canadian mining companies to provide assistance to those affected by their operations. In

---

20 Rights and Democracy's most recent efforts in Latin America were already suggesting a subtle realignment. As president, Latulippe spoke out against 'democratic backsliding' in Venezuela, which he criticized for seeking to restrict freedom of association through a proposed law that would disallow Venezuelan organizations from receiving foreign funding (Berthiaume 2010b). The organization also granted its annual democracy award to a Venezuelan dissident, and initiated new programming in Colombia in support of human rights groups making claims through the judicial system (Rights and Democracy 2011).

September 2011, CIDA (2011a) announced $26.7 million in grants to WUSC, Plan Canada, and World Vision Canada for projects they will conduct in conjunction with mining giants Rio Tinto Alcan, IAMGOLD, and Barrick Gold, respectively. As Schulman and Nieto (2011) note, the new partnership means that 'the Canadian government is able to deflect demands for more stringent – and potentially profit-damaging – controls over one of its most lucrative industries'. Indeed, the announcement was made in the context of the new Andean Regional Initiative for Promoting Effective Corporate Social Responsibility, which promotes voluntary over binding standards in Peru, Bolivia, and Colombia. The use of Canadian funds to contain the operational fallout of billion-dollar companies represents a disconcerting and dangerous trend in Canadian development assistance. It is also likely to drive a wedge between those NGOs willing to align their activities with the state and the interests of Canadian multinationals and those who insist on a more principled stance.

Concluding Remarks

The new millennium has thus witnessed changes in North American democracy promotion practices with important underlying continuities. For the United States, democracy promotion remains inextricably linked to geopolitical and commercial objectives. This is also the case for Canada, which has developed a more strategic approach in keeping with the reorientation of Canadian foreign policy that began more than two decades ago.

Shifts in the regional order that emerged at the end of the twentieth century, however, have conditioned the use of democracy promotion as a modality of power. With the popular mobilization against neoliberalism, many countries have swung to the left while others have contained social dissent through a strategy of passive revolution. During the Bush administration, a strategy of regime change was pursued to reverse the threat posed by the radical left in Haiti, Venezuela and Bolivia; in the case of Haiti, the United States received the support of the Liberal government in Canada. At the same time, most democracy assistance through USAID actually went to stabilize allies to contain the leftist threat. With the international and regional backlash against democracy promotion, strategies of regime change faced new constraints, and stability and security have become the dominant political objectives of democracy promotion for both the Obama administration and the Harper government.

If the new global conjuncture has given rise to a backlash against US democracy promotion, the efforts of the Canadian state to rationalize and centralize its approach to democracy promotion offers new possibilities to provide low-intensity democracy with a universalistic makeover. Canada seems to have eagerly adopted this role, redefining the way in which democracy is conceptualized by Canadian practitioners and undertaking considerable efforts to reorganize the democracy promotion field of practice under the leadership of DFAIT and CIDA. Canadian middlepowermanship, in short, has increasingly come to define Canadian democracy promotion as a field of practice in the service of neoliberal governance. In such circumstances, it is unlikely that the rights-based or grassroots tradition will survive the institutional and discursive changes underway; indeed, the process of replacing it is well underway

As we shift the analysis to the case studies, I will explore in more detail how these macro developments in North American democracy promotion and regional order have played out in the context of specific national settings. There are three key themes that will be explored in detail: 1) the intersection of US and Canadian strategies of regime change in Haiti and the shift to stabilization following the 2004 coup; 2) the use of democracy promotion in Peru by both Canada and the United States to reinforce the state's strategy of passive revolution; and 3) the shifting nature of democracy promotion in Bolivia and the backlash that ultimately forced the United States to abandon its strategy of regime change. Differences between the approaches of Canada and the United States – particularly in Bolivia and Peru, where the positive effects of the grassroots approach were still apparent throughout the period in question – will be explored. In looking at these themes, greater theoretical clarity to the framework that has been developed thus far will be provided by exploring the mechanisms, practices and discourses through which democracy promotion contributes to neoliberal hegemony.

Chapter 3
Polyarchy at any Cost in Haiti

Two and a half decades after the fall of the Duvalier dictatorship, the consolidation of democracy in Haiti has been thwarted by coups, flawed elections, and intermittent bouts of repression. According to many development experts, Haiti is the quintessential failed state, an impoverished nation whose very existence depends upon the goodwill of international donors. Yet few acknowledge that many of Haiti's international partners have contributed to the process of state failure, forcing upon it economic reforms that have decimated its productive base and weakening its political institutions by meddling in its social conflicts. Haiti's fractious elite, unwilling to produce a more democratic order based on a fairer distribution of resources, has long relied upon US assistance to maintain its dominant status through interventions that have reinforced the country's extreme dependency. In the last decade, however, Canada has contributed to the US project of imposing a form of low-intensity democracy in Haiti through its own democracy promotion efforts. As the Bush administration launched a new round of democracy wars in the Americas to roll back leftist governments – and the Canadian state began to align itself more closely with Washington's agenda – the small island republic became a central front in the conflicts that would mark the first decade of the new millennium. The government of Jean-Bertrand Aristide became the first major casualty of the hemispheric struggle. Haiti, where North American influence has been most overtly political, provides an important case study in the contradictory politics of regime change through democracy assistance.

The key turning point in the analysis that follows is the winter of 2004, when President Jean-Bertrand Aristide was forced to flee the country for a second time as paramilitary factions controlling large sections of the countryside made their way to the capital to depose the government. Those who drew parallels with the bloody coup that had overthrown Aristide soon after Haiti's first-ever democratic elections were quickly told that the events of 2004 were nothing like those of 1990. This time around, civil society was unanimous in its insistence that Aristide had squandered his popular mandate through corruption and human rights abuses, generating a crisis of authority that could only be resolved through his immediate resignation. In fact, Aristide was neither saint nor devil – despite the government's reliance on organized gangs in the slums of the capital to defend it from opposition forces, it maintained popular support. Yet the narrative of the priest turned dictator provided a pretext for Canadian and US programs that helped reinforce the hand of traditional elites who were uninterested in resolving the conflict through democratic means.

This chapter will examine who has benefited from North American programmes, and how they have respond to the evolving balance of power in Haitian civil society and the state before and after the coup. It will assess what factors motivated Canada from adopting an aggressively interventionist approach in Haiti, when its democracy assistance programs had never been used before so flagrantly to undermine a government. The analysis is divided into four parts. I begin by setting the historical context through a short description of Haiti's social structure of accumulation, its commercial relations with the United States and Canada, and the controversies surrounding the second presidency of Jean-Bertrand Aristide (2000–2004). The second part examines North American democracy promotion during this period, when Ottawa and Washington responded to the threat posed by Aristide and the Lavalas movement to traditional class relations through a strategy of regime change. Although Aristide had implemented most of the neoliberal reforms demanded by the international financial institutions, he had done so begrudgingly, and the local elite and their Republican allies in Washington remained viscerally opposed to him and the popular movement. With growing material interests in Haiti and considerable cultural capital, the Liberal government in Canada aligned itself with Washington's agenda. Through USAID funds, IRI helped build a coalition of opposition parties while a lesser-known US democracy promotion agency, IFES, mobilized civil society opposition groups. The leaders of the opposition fronts were deeply tied to transnational capital and shared common interests with Canada and the United States. CIDA contributed to the campaign to destabilize Aristide's government by channelling funding to organizations and coalitions representing Haiti's small middle class. Such funding helped distort the perception of the social basis of opposition to Aristide.

A key question that this chapter will seek to answer is why Canadian NGOs joined the campaign against Aristide. For if Canada's unofficial opposition to Aristide was motivated by growing material interests in Haiti and the development of a more muscular middlepowermanship, Canadian NGOs were not merely carrying out the state's agenda. Most had in fact internalized the structural split in Haitian civil society between middle-class and elite sectors on the one hand and more popular sectors on the other. In a context of extreme polarization, they aligned themselves with their counterpart Haitian NGOs, who were largely disillusioned with the Lavalas government and remained detached from the popular movement. This helps account for the differences that we will observe in the behaviour of Canadian NGOs in Peru and Bolivia, where the relationship between NGOs and popular movements has been less polarized and Canadian NGOs have taken on a more progressive character (though, as we will see, this is changing). But it also reflects the intrinsic weakness of the grassroots approach, which can be mobilized towards unpopular ends.

The third part of the analysis focuses on democracy promotion under the interim government of Prime Minister Gérard Latortue (2004–2006), during which time American and Canadian democracy assistance programmes continued to support elite social forces as the government repressed the popular movement. Despite this repression, Aristide's former Prime Minister, René Préval, won the presidential elections of 2006. With the destruction of the Lavalas base, however, Préval rose to power at the head of a weak political coalition which accepted the neoliberal agenda. Following his victory, democracy assistance programmes have placed greater emphasis on building the legitimacy of the state to stabilize the political situation. Although I refer briefly to some of these programmes, the tactics of stabilization are not theorized in depth until the following chapter on Peru.

The chapter ends with some comments on the political situation following the terrible earthquake of January 2010. Although Canada and the United States have delivered badly needed aid, they have supported elections marked by significant irregularities and the ongoing exclusion of Lavalas from political life. In the first round presidential election of November 2011, only 22.8% of the electorate showed up at the polls. Michel Martelly – a popular singer with close ties to the most reactionary sectors of Haiti's elite – emerged the lucky winner of the second. If a polyarchy seems finally have to emerged in Haiti under the leadership of the elite, it has done so at the expense of popular sovereignty.

Elite Civil Society against the State: Aristide's Second Presidency

Haiti has endured a lengthy history of endemic poverty, inequality and subordination in the world system. The vast majority of Haiti's population is of African descent, with a small mixed African-European, or *mûlatre*, elite comprising only 5% of the population (Central Intelligence Agency 2012). Based in the urban centres and inhabiting an exclusive cultural world alongside a small black political elite, the *mûlatre* speak French, the language of the former colonial rulers. Creole, a French patois with strong African, Spanish and English influences, is spoken by Haitians of all social classes. The wealth of the dominant class is linked to assembly manufacturing and agro-exports; the vast majority, however, work the land with little access to vital inputs while the slums of the nation's capital overfill with displaced peasants and unemployed workers in some of the most difficult conditions of the Americas.

Crisscrossed by mountains and valleys on the eastern side of the island shared with the Dominican Republic, Haiti has suffered from one of the world's most extreme cases of overpopulation and deforestation on the earth – a fate which has not been compensated for by any significant natural-resource base. Although recent statistics on inequality and poverty are unavailable, 65% of all Haitians lived below the poverty line a little over a decade ago before the devastating effects of the earthquake; the richest 10% of Haitians held nearly 50% of all income while the poorest 10% held a paltry 0.7% (UNDP 2007).

The United States has historically enjoyed a substantial trade surplus with Haiti, which has one of the most liberalized tariff structures in the world. In 2000, US exports to Haiti equalled $576.7 million while imports were $296.9 million; by 2008, exports had increased considerably to $944 million while imports also rose to $450 million (US Census Bureau 2010a). Canadian exports to Haiti equalled approximately $51.4 million and imports $19.2 million (Government of Canada 2009a). Although both Canada and the United States have considerable investment in Haiti – the only two foreign banks operating in Haiti are Citibank and Scotiabank of Canada – the Haitian government does not keep figures on FDI breakdown by country (total FDI inflows in 2008 amounted to $19.3 million). Major US companies operating in Haiti include Texaco (Chevron), Seaboard Marine, Continental Grain, Trilogy Inc., Spirit Airlines and Newmont Mining (US Department of State 2009). Major Canadian investors include SNC-Lavalin, Gildan Activewear, Bank of Nova Scotia, and S.M. Group International Inc. (Grant 2011).

Haiti's subordinate position in world order and its special relationship to the United States have historically shaped its class and ethnic relations. The social structure that crystallized following the world's only successful slave revolution in 1804 was defined by an alliance between a new black political-military elite and a *mûlatre* merchant class at the expense of the peasantry. Within the new formation, the state took on a parasitic form in which political office served the primary function of economic accumulation (Dupuy 1997). Political instability in the early twentieth century served as the pretext for the invasion of US Marines in 1915, which led to an occupation that lasted until 1934 and strengthened the hand of the *mûlatre*. Under the dictatorship of Francois Duvalier (1957–1971), however, the black political elite reasserted its privileged position. Although the United States was weary of Duvalier's nationalist rhetoric, the Anti-Communist Law passed in 1969 set the stage for a rapprochement. When the reigns of the dictatorship passed to Jean-Claude Duvalier (1971–1986), USAID oversaw the restructuring of Haiti's economy through low-wage assembly manufacturing for the US market, a process which further strengthened the commercial elite (Dupuy 1997).

By the early 1980s, a popular civil society opposition rooted in the urban poor and the peasantry emerged under liberation theologian Father Jean-Bertrand Aristide's charismatic leadership. When presidential elections were held in 1990, the United States provided democracy assistance to liberal factions of the elite opposed to the popular movement (Robinson 1996). Aristide easily won the contest with 67.5% of the vote at the head of the Lavalas movement (Creole for avalanche), before being deposed by a military coup mine months later. By 1994, however, the Clinton administration decided to reinstate Aristide on several conditions (see Clement 1997), including the acceptance of an Emergency Economic Recovery Program formulated by the World Bank, IMF, Inter-American Development Bank and USAID with the support of CIDA and several other donors (United Nations 1995), which called for the mass privatization of Haitian state enterprises, which were deemed corrupt and inefficient. While this was undoubtedly true, Aristide had advocated turning them into cooperatives to benefit the masses rather than a small group of private interests (Hallward 2007). The international plan was thus a direct violation of the modest political and economic vision for which the Haitian population had voted for prior to the coup.[1]

Canada, which began providing development assistance to Haiti in the late 1960s and refuge to a burgeoning diaspora fleeing the dictatorship throughout the 1970s and 1980s, took on a more active role in Haiti following Aristide's reinstatement. Under the Liberal government of Prime Minister Jean Chrétien, Canada played a major role in sustaining a UN peacekeeping operation financially and with personnel. Although Canada unequivocally supported Aristide, its commitment to democracy was attenuated by its support for the political and economic programme that was a condition of his return, and which systematically strengthened elite sectors of civil society against the popular movement which had been the driving force behind the democratization process (Shamsie 2007).

Aristide was succeeded by his prime minister, René Préval, in the 1996 presidential election. In the 2000 general election, *Fanmi Lavalas*[2] achieved a majority in the Chamber of Deputies and Senate, with Aristide returning to the presidency soon after. Aristide won by a landslide, but most opposition parties had refused to field a candidate after claiming fraud in the general election.[3] In fact, the opposition remained marginal as Lavalas continued to enjoy high levels of popular support two years into Aristide's presidency despite the worsening economy and pressure on Aristide (see Table 3.1). At the same time, however, many of Haiti's most significant peasant organizations, intellectuals and other supporters condemned Lavalas for its cooperation with structural adjustment and accused it of becoming *anti-populaire* (Hallward 2007).

---

1   Had Haiti's donors been serious about alleviating endemic poverty, they might have, as Shamsie (2007) emphasizes, encouraged a sustainable agricultural policy that would have funnelled resources to the peasantry to enhance local subsistence production. Lavalas itself had called for increased support to the peasantry in the form of micro-credit, state-led cooperatives, extended irrigation and land reform to improve the lot of the peasantry (Dupuy 1997).

2   The successor to the original Lavalas after the party split in 1995, primarily over the privatization issue. I use the terms FL and Lavalas interchangeably.

3   The senatorial elections were called into question when the *Conseil Electoral Provisoire* (CEP) applied a vote-apportioning methodology to grant first-round victories to eight senatorial seats won by FL candidates who had won considerable majorities but had not obtained 50% of the vote. The same methodology had been used in the 1990 and 1995 elections (Hallward 2007).

Table 3.1 Poll on political preferences in Haiti (2002)

Political Preferences	Results	Social sector
Adult population identifying with FL Level of support for Aristide	33% 50%	High level of support among women and the poor
Level of support for opposition	9%	Support concentrated in middle and upper class

Source: Dupuy (2005).

In the years that followed, the opposition declared that Aristide was dangerously backsliding into authoritarianism, committing massive human rights violations and encouraging wide-scale state corruption. Peter Hallward (2007) has refuted these claims, demonstrating how much of the violence that occurred can be attributed to the *Police Nationale d'Haiti* (Haitian National Police – PNH), whose political affiliation was often anti-government, or to pro-FL groups rather than the government itself.[4] On the issue of corruption, there is no question that many Lavalas cadres were engaging in rampant corruption and drug trafficking, though this was nowhere near the official state-sanctioned piracy of the Duvalier years or the military dictatorship of the early 1990s. Such patterns of accumulation, moreover, were deeply rooted in the political economy of Haiti and the parasitic state form that Aristide inherited.

Canada and the United States officially responded to the crisis by supporting a compromise solution. In reality, however, both undermined the government. Although Aristide had acquiesced to most political and economic demands, the Haitian elite remained viscerally opposed to Lavalas and its redistributive agenda (however modest) (Fatton 2002). Republicans remained closely aligned with the elite and shared its ideological hatred of Aristide and the Lavalas movement. Canada likely saw an opportunity to strengthen its economic interest in the country while enhancing its diplomatic stature on the regional scene. Its active involvement in Haiti, moreover, provided an opportunity for rapprochement with

---

4   Michael Deibert (2005), whose book *Notes from the Last Testament* documents the link between the Lavalas government and armed gangs, provided a detailed criticism of Hallward's work for being partisan and factually inaccurate.Although Hallward's (2008) rebuttal provides a convincing point-by-point refutation of the main criticisms, Hallward does indeed fail to sufficiently problematize Aristide's reliance upon armed gangs to break up opposition rallies. Dupuy (2005) provides a more nuanced critique of Aristide's second term that takes into consideration the role of the dominant class and international forces in undermining democracy in Haiti, but is scathing in its analysis of Aristide's acceptance of structural adjustment and his failure to mobilize popular forces behind a progressive alternative. The recourse to the gangs, in effect, symbolized the inability of the government to arouse the people to its defence. While I do not dispute Dupuy's interpretation, the limitations of Lavalas certainly did not justify the international onslaught against it.

the Bush administration after the decision of the Chrétien government not to follow the United States into war in Iraq. Both the United States and Canada thus worked to undermine the political forces that threatened the class power and state form which underpinned Haiti's position in neoliberal regional order.

George W. Bush's administration led the way by suspending all aid after the disputed elections. With the International Financial Institutions (IFIs) following suit, the aid embargo significantly weakened the government with disastrous effects on social policies.[5] Under the Liberal governments of Jean Chrétien and Paul Martin, Canada was one of the few donors that maintained aid to Haiti, dispersing over $18 million from 2000 to 2001. But this sum actually represented half the amount that was disbursed the previous fiscal year and most of it was redirected from the state to NGOs. If Haiti thought it had an ally in the Canadian government, it was badly mistaken; in 2003, the Secretary of State for Latin America, Africa and the Francophonie, Denis Paradis, organized a secret meeting of states and international organizations concerned with Haiti at Meech Lake which determined that Aristide should be removed from power and Haiti placed under UN administration under the Responsibility to Protect doctrine (Engler and Fenton 2005). Meanwhile, as rightwing paramilitary factions began to take over the central plateau region, Secretary of State Colin Powell dismissed Aristide's popular support base, calling on him to take 'action to reach out to the opposition to make sure that thugs are not allowed to break up peaceful demonstrations'. Canadian Minister of Foreign Affairs, Bill Graham noted further 'that we are frustrated with the violence, we are frustrated with Aristide' (CBC News 2004).

With the paramilitaries advancing on the capital, Aristide was forced to board a plane by private US security forces (see Hallward 2007). Speaking before the House of Commons, Bill Graham claimed that it was his decision to leave, and one that likely 'spared his nation worse violence, indeed the possibility of a humanitarian catastrophe'. Liberal MP David Kilgour rose in the House to assert that the opposition parties were not linked to the rebel forces and had no control over their actions. 'Haiti is a failed state,' he asserted, 'tragically where anarchy and chaos reign, and the rule of law is non-existent' (House of Commons 2004).[6] With Aristide gone, a Multinational Interim Force (MIF) composed of troops from Canada, the United States and France deployed to stabilize the situation. Soon after, a UN peacekeeping mission led by Brazil and Chile, the *Mission des Nations Unies pour la Stabilisation en Haiti* (United Nations Mission

---

5   The Aristide government continued to pay arrears on its debt to the Inter-American Development Bank (most of it odious) despite the fact that the IDB caved to pressure from Washington and froze four loans approved in 2000. In July 2003, Haiti sent more than 90% of its foreign reserves to the IDB to pay off these arrears (Farmer 2004).

6   The ties between the *Convergence Démocratique* and the paramilitaries, however, were made clear through multiple public statements by paramilitary leaders (Hallward 2007).

for Stabilization in Haiti – MINUSTAH), relieved the MIF of its mandate. With its experience in waging war against the poor of the slums of its megacities, the Brazilian military was particularly well placed to lead the mission.[7] The timing of the mission leaves little doubt that Haiti's partners were only interested in restoring stability with Lavalas out of power.[8] While several states contributed troops to the mission, the Organization of African Unity (OAU) and the Caribbean Community and Common Market (CARICOM) demanded an investigation into the events surrounding Aristide's departure, refusing to recognize the legitimacy of the interim government that took its place.

Haiti had come full circle and an external intervention was once again required to break the class stalemate resulting from its polarized social structure. As the government headed by Prime Minister Gérard Latortue began implementing a new neoliberal plan approved by the international community, little question as to whose interests the coup government represented remained. Although the victory of René Préval in 2006 marked the restoration of democratic rule, the decimation of the popular movement ensured that the new president would not budge from the strait-jacket imposed by the international presence.

Creating a Crisis of Authority

The American QUANGOs Take the Lead

In the years leading up to the coup, both CIDA and USAID expended considerable funds supporting elite civil society, though both spent considerably more on stabilization efforts once Aristide had been forced to flee, as depicted in Figure 3.1. From 2001 to 2004, CIDA spent $11.5 million in democracy assistance compared to $81 million over the course of the next three years. Likewise, the United States spent $13.8 million from 2001 to 2003 and $66.2 million over the next three years (to put this into perspective, the cash-starved government of Aristide was operating on a budget of $300 million in 2003 (Farmer 2004)). According to a

---

7   The role of the rising Latin American powers in Haiti raises the issue of regional sub-imperialism, a topic that warrants considerable more attention than that which can be provided here. One possibility is that Haiti's historical subordination has rendered it an unfortunate victim of imperial and sub-imperial aggression as different sub-powers have sought to establish credibility and prestige through the ongoing intervention. Although their interests have roughly coincided, this should not be taken to mean that the Latin American and North American states share a common vision of regional order. Certainly, the reaction of Brazil and others to US intervention closer to home speaks to the distinct spheres of influence now being carved out; Haiti's unlucky predicament is to find itself at the geographic and historical interstices of those spheres.

8   As Shamsie (2006) states, the point is not that Canada should have acted to protect Aristide the man, but rather that it 'should have done much more within its power to support his office and position as a constitutionally-elected leader'.

CIDA official in Port-au-Prince, Canada and the US coordinated most of their democracy promotion efforts following the controversial 2000 elections, and there was little difference between them.[9]

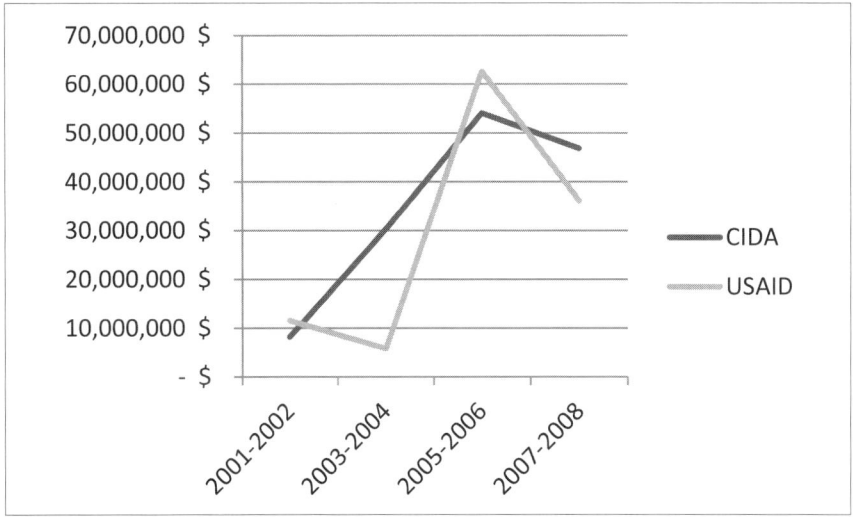

Figure 3.1 US and Canadian democracy assistance before and after the coup

But if the United States and Canada both mobilized elite social forces against Aristide's government, they did not use the same tactics, nor did they target the same actors. Elite civil society was composed of different sectors and a certain informal division of labour between Canada and the United States occurred in supporting different groups in the overall interest of creating a larger hegemonic bloc. This bloc was led by transnationalized fractions but also included elite social groups and NGOs critical of neoliberalism.

USAID's $12.2 million programme introduced in 1999, More Genuinely Inclusive Democratic Governance, provided the framework through which various American QUANGOs channelled US resources to coordinate the elite opposition. These efforts were undertaken chiefly by IRI and the International Foundation for Electoral Systems (IFES). The programme purported to work with 'organizations in all development sectors to build their advocacy skills so that they may positively influence government policies and oversee public institutions' (USAID 2000).

IRI in particular had long established a pattern of meddling in Haitian politics against the interests of the Lavalas government. After having been forced to close its office in Haiti in the late 1990s, IRI's fortune changed with the return

9   Interview. Port-au-Prince: 12 March 2009.

of Republican hardliners in Washington, including far-right operatives in the US Department of State such as Otto Reich and Roger Noriega, both veterans of the first round of democracy wars against the Sandinistas in Nicaragua. These key figures in the Republican administration provided unconditional support to the head of the IRI programme in Haiti, Haitian-American Stanley Lucas, who played an instrumental role in uniting the disparate factions that formed the new political opposition to Aristide, the *Convergence Démocratique* (Democratic Convergence).

From 2002 to 2007, IRI received $4.3 million in federal funds to manage its training programmes for civic education and civil society organizations and political parties, which did not include a single member from the most popular party in Haiti. Training sessions typically lasted two days and included sections on a number of political strategies, from developing issue-based political party platforms to encouraging political coalition-building (IRI 2008b). According to a *New York Times* investigation, US Ambassador Brian Dean Curran – a Clinton appointee – repeatedly expressed his concerns about IRI's activities to his superiors in Washington in the Bush administration's Latin America policy team (Bogdanich and Nordberg 2006). The agents of democracy promotion thus became the vehicle through which an informal foreign policy was delivered, providing the government with a shield of deniability from possible accusations that it was fomenting a coup.

The partisan activities of IRI contrasted with the more balanced approach of NDI. A total of 10 national political parties were supported through its political party strengthening programme, including Lavalas, six parties associated with the CD, and three non-aligned parties.[10] Despite its less partisan approach, however, the NDI programme served to further distort the political landscape by building parties which lacked popular appeal on their own terms.

IFES programmes in the early 2000s focused on empowering civil society organizations based on the rationale that the executive controlled the judiciary and that the government was deeply implicated in human rights abuses (IFES 2004). The creation of several justice-focused organizations and their 'sensitization' was instrumental in forming the basis of the civil society opposition, the so-called *Groupe des 184* (G-184). A University of Miami report (Griffin 2004: 22) noted that interviewed IFES administrators went so far as to claim that they would have liked to have taken full credit for Aristide's eventual ouster but could not out of respect for the US government (Griffin 2004).

The democracy promotion projects implemented by IFES in the early 2000s were organized under the titles Civil Society Strengthening Project for Judicial Independence and Justice and Victims of Organized Violence Program.[11] Both programmes were premised on the assumption that the executive controlled the judiciary and that the government was deeply implicated in human rights abuses. This, in conjunction with a congressional ban on funding for work with the

---

10   Interview with NDI Resident Director. Port-au-Prince: 15 July 2008.

11   The budget size for these projects was not publically disclosed; despite repeated attempts, IFES did not respond to interview requests.

government of Haiti in the wake of the 2000 elections, provided the justification for the unorthodox focus on civil society for a justice-reform project. Among other things, IFES set up an umbrella organization for the legal sector, worked with journalist groups and radio stations to publicize Aristide's alleged corruption and formed a student group, *la Fédération des Étudiants Universitaires d'Haïti* (Federation of Haitian University Students – FEUH), at the state university in Port-au-Prince (Griffin 2004).

With funding from IFES as well as other donors, the G-184 – whose leadership came from business sectors closely integrated with transnational capital – sought to counter Lavalas' popularity by unveiling a new 'Social Contract', which gave the appearance of a coherent political alternative. Under the leadership of sweatshop magnate, Andy Apaid, the G-184 launched a national caravan to build support for the Contract across the country. Although the turnout for the caravan was mostly weak, the initiative symbolized the effort to disseminate the anti-Aristide discourse and build a popular hegemony to legitimize the opposition.

IFES also played an important role in manufacturing a human rights crisis that significantly tarnished the image of the Lavalas government. The most controversial element of IFES programming was the funding of a human rights hotline administered by *le Comité des Avocats pour le Respect des Libertés Individuelles* (Committee of Lawyers for the Respect of Individual Rights – CARLI). IFES provided the group with $54,000 to create the hotline and provide monthly written reports detailing alleged abuses and their perpetrators to radio stations, the US Embassy, the OAS, and various domestic and international organizations. Once IFES funding came to an end during the reign of the interim government, however, CARLI admitted to the human rights delegation of the University of Miami that it was under a lot of pressure by IFES to denounce Aristide, having gone so far as to publish names of perpetrators with little or no investigation (Griffin 2004: 28). The National Lawyers Guild (2004) reported that those who were condemned by the hotline were never contacted to respond to the allegations and that CARLI, with no full time staff and only two volunteer lawyers at its office, hardly had the capacity to investigate the 60 to 100 calls it was receiving a month.[12]

US policymakers and intellectuals on the Right also realized the importance of reinforcing this perception at home to ensure that Democratic appeals to defend Aristide were marginalized in policy circles. To this end, the Haiti Democracy Project, a Washington-based research group which brought together former US ambassadors and influential members of the far-Right Haitian-American community such as Lionel Delatour from the Anti-Aristide group, *Centre pour la Libre Entreprise et la Démocratie* (Centre for Free Enterprise and Democracy – CLED), was launched in November 2002. The project provided another illustration

---

12 The National Lawyers Guild has been an outspoken critic of US democracy promotion efforts. See the Guild's *Resolution on the Misuse of U.S. Government "Democracy Promotion" Initiatives Undermining Progressive Governments and Movements in the Americas* at: http://www.nlginternational.org/news/article.php?nid=44.

of the proliferation of 'democracy' groups representing the same social sectors through an extended network of interlocking elite personnel. As Aristide came under siege, the project sent a delegation of former US ambassadors to Capitol Hill to advocate that 'the transition be allowed to run its course' (Haiti Democracy Project 2007). American democracy promotion may not have succeed at welding together a popular hegemony, but it did manage to coordinate the efforts of numerous elite in Haiti and the United States to trigger a crisis of authority that made it easier for the most reactionary elements of Haitian society to deliver the *coûp de grace* against Aristide's government.

The recourse to force exposes the central political myth of this period – that civil society was unanimous in its opposition to Aristide. The credibility of this argument has always depended upon the erasure of the agency of Haiti's majority. Indeed, as Haitians in the popular movement argue, Haiti's civil society can be distinguished between its minority and majority component (*société civile minoritaire* versus *société civile majoritaire*).[13] In the former camp stood the opposition, including many progressive NGOs which received support from Canada (as discussed below); in the latter camp, there were the *organisations populaires* (popular organizations, also referred to as the *organisations de base*) which comprised the Lavalas movement, as well as most of organized labour.

The largest labour federation, for instance, the *Confédération des Travailleurs Haïtiens* (Confederation of Haitian Workers – CTH), supported the Lavalas government until the very end. According to a delegation with which I met, the federation represents over 65,000 members – a considerably larger social base than all of the opposition organizations combined. While its leaders were not uncritical of the decisions of the Lavalas government, they insisted on the right of President Aristide to complete his term in office. The OPs themselves number in the hundreds, providing various social services to the local population. They overlap with hundreds of small groups (*ti fanmis*) that occasionally mobilize at the regional level, and whose political programme revolves around the three main themes that have long informed the Lavalas movement – justice, transparency, and participation. There is no question that they are organizationally weak and often unable to mobilize politically beyond the short term – factors which reflect the extreme precariousness of the class status of their leaders.

Commenting on the transnational alliance that undermined the Lavalas government, one activist put the matter succinctly: 'There was a minority civil society composed of the bourgeoisie, the political class and of course some organic intellectuals who always defend this privileged minority. And – you may laugh but it is true – one also has to place within this civil society the United States and its allies. I'm speaking of the government of George Bush, the government of Canada, and the government of Chirac, all of which supported the coup.'

---

13   Interview with delegation from *Réseau d'Organs National Multiplicateur de Fanmi Lavalas*, Port-au-Prince: 16 March 2009.

It should be noted that some non-state funded US NGOs have partnered with the popular movement from an approach that emphasizes solidarity and accompaniment such as the Institute for Justice and Democracy in Haiti (IJDH) and Partners in Health (PIH). Although PIH has not worked directly on the issues of democracy and human rights, its approach to development in Haiti is notable since it has sought to empower local community organizations such as the OPs from a perspective that emphasizes the link between sickness and socio-economic inequality. IJDH's activities have been more explicitly political, and its activism provides a good example of the counter-hegemonic potential of transnational linkages across the region. The US-based NGO partners with the public-interest law office, the *Bureau des Avocats Internationaux* (Bureau of International Lawyers – BAI), an organization which works directly with the OPs assisting victims of human rights violations receive legal compensation and defending those who have been falsely accused.

The Canadian Contribution

Despite the ongoing popularity of the Lavalas government and its organized social base, Canada chose to reinforce the minority civil society. CIDA added value to the destabilization campaign, however, by targeting funding at small organizations and coalitions that represented Haiti's small middle class, including some which were nominally on the left but opposed to the popular movement. Such funding, much of which was administered by Canadian NGOs, helped to distort the perception of the social basis of the opposition to Aristide.

CIDA's direct contribution to this distortion was the funding of women's organizations that represented themselves as the vanguard of a cross-sector opposition to Aristide. Between 2000 and 2004, CIDA provided funding to two women's organizations, the *Coordination Nationale de Plaidoyer pour les Droits des femmes* (National Coordinator for the Advocacy of Women's Rights – CONAP) and EnfoFanm (Women's Information), both of which issued regular statements denouncing alleged-human rights violations being committed by Aristide's government and calling for its immediate resignation. CIDA also provided funding to other women's organizations which opposed the government including *Kay Fanm* (Women's House), *Solidarité Femmes Haïtiennes* (Haitian Women's Solidarity – SOFA) and *Fanm Deside* (Women Decide) through the Canadian NGO, Development and Peace, and, in the aftermath of the coup, continued to fund CONAP though a project implemented by Rights and Democracy.

Although such organizations have engaged in important initiatives denouncing violence against women and advocating for women's rights, their leaders are French-speaking development professionals removed from the day-to-day struggles of the impoverished masses. This in itself does not discredit their actions, but the fact that they do not organize amongst the people nor show solidarity with the popular sectors of Haitian society in their political struggle with the dominant class demonstrates the divide between many NGO feminists and the class struggle in which poor women from the *organisations populaires* are engaged.[14]

Such organizations were particularly critical of the Lavalas government's support to the so-called *chimères* (Creole for phantoms) – gang members affiliated with Lavalas who were allegedly responsible for using rape as a political weapon against opponents of Aristide. This is a claim that is often repeated with little documented support. In fact, one of the more visible leaders of CONAP and ENFOFANM who carried out research on the political use of rape since the Duvalier years admitted as much in an interview. 'The findings,' she stated 'were that civil rape was vastly more important than rape in the political sense. This is not what we are supposed to say but I cannot deform the reality. The other rapes are more spectacular but no one has been able to show me that political violence toward women was higher during the Aristide years.'[15]

CIDA funds to Aristide opposition groups were also channelled through a handful of left-leaning NGOs such as Development and Peace, Alternatives, and the QUANGO Rights and Democracy. The most active of these organizations, Development and Peace, whose Haiti Country Program ran from 2003 to 2006 with a total budget of 1.1$ million, played a role similar to IFES in mobilizing civil society opposition groups against Aristide's government. As a programme officer for Haiti explained, the NGO's activities in the lead up to Aristide's 'departure' were informed by the perspective that Aristide was arming Lavalas supporters who were committing human rights violations and repressing civil society. Specifically, it sought to strengthen over 70 'popular grassroots organizations' to make them 'more democratic and more autonomous', including members of the G-184 representing seemingly more progressive constituencies in the women's and peasant's movements, such as *Kay Fanm*, SOFA, *Fanm Deside* and the *Mouvement Paysan Papaye* (Peasant Movement of Papye – MPP). Development and Peace justified supporting these sectors by reducing Lavalas to the *chimères* and dismissing – pejoratively and ethnocentrically – its popular support base.

Despite the assertion that it was supporting the poorest sectors of society in a struggle against authoritarianism (Development and Peace 2006), the groups funded by Development and Peace possessed tiny membership bases and centred primarily on the personality of one dominant individual. One of the few G-184 organizations with a substantial, though purely regional, following, was the MPP. Over the years, its leader, Chavannes Jean-Baptiste, had managed to build a substantial peasant following in Hinche, sustained in large part by the steady

---

14 Women activists in the Caribbean acknowledged the gap between feminists and the popular movement when one group issued a statement against the coup in March 2004. Peggy Antrobus (2004), former General Coordinator of Development Alternatives with Women for a New Era (DAWN) and signatory to the statement, noted that the women's movement in Haiti suffered the same class divisions of women's movements everywhere, and that it has suffered from co-optation.

15 Interview, Port-au-Prince: 18 March 2009.

flow of NGO patronage that passed through the MPP's hands. Its grassroots activism, however, remained easily co-opted by its dominant leader. In the 2006 presidential elections, Jean-Baptiste was able to use the MPP as a political vehicle to mobilize a support base for one of the most reactionary G-184 leaders, the far-right industrialist, Charles Baker, with whom he had created a close political alliance (Baker only received 8.24% of all votes) (Hallward 2007: 104).

The Montreal-based NGO, *Alternatives*, and Rights and Democracy played a similar role in demonizing Aristide as the Haiti Democracy Project in the US. The *Alternatives* web-site regularly published articles by *Alterpresse*, a Haitian NGO that served as one of the main vehicles for the dissemination of anti-Lavalas propaganda. Although Rights and Democracy did not open an office in Haiti until after the coup, it marshalled its own credibility in Canada to discredit the Lavalas government. A report published in 2004 echoed many of the arguments advanced by IFES, depicting Aristide as a corrupt and intransigent ruler who repeatedly refused to make concessions to his opponents and was largely responsible for the country's political impasse. The G-184 was portrayed as representative of 'various sectors of civil society' 'promoting a new model of governance in the form of a social contract'. Nowhere is it mentioned that the opposition was in fact headed by industrialists such as Andy Apaid, Charles Baker, and Reginald Boulos, all of whom were outspoken advocates of greater privatization and the establishment of Export Processing Zones (for which Aristide is duplicitously criticized; the report failed to mention that Aristide himself had always been a vocal opponent of neoliberalism and that the international community had regularly frozen or stalled the flow of aid to ensure the implementation of neoliberal policies).

The involvement of NGOs in the opposition against Aristide both in Canada and in Haiti presents a paradox – why would nominally left organizations both in Canada and Haiti side with a US-funded elite in its campaign to destabilize a popular government? In the months leading up to Aristide's forced departure, many Québécois NGOs even created an informal network to coordinate their actions, the *Concertation pour Haiti* (Concertation for Haiti – CPH), which brought together a dozen or so labour organizations and NGOs, including Development and Peace and Rights and Democracy. The answer cannot be reduced to a single factor. Within Haiti, many of the most vocal opponents of the Lavalas government were former militants frustrated by their exclusion from high office. Discontented petty-bourgeois leaders increasingly turned to marginal NGOs and political parties in the late 1990s and early 2000s as vehicles to advance their own ambitions. Transnational penetration and the 'NGO-ization' of Haitian civil society (Schuller 2007) facilitated this process through the proliferation of thousands of NGOs. The dependence of local organizations on foreign funding did not necessarily reflect imperialist co-optation, however, but signified the establishment of a mutually beneficial arrangement. But certainly not all of those who opposed Aristide were acting out of a clearly-defined conception of self-interest and opportunistic calculation. Some raised legitimate critiques of Lavalas, particularly surrounding its failure to mobilize different sectors of the population against the neoliberal onslaught. In systematically denouncing the actions of a popularly-supported government, however, such NGOs opened the way for highly reactionary forces to return to power and pursue an openly unpopular programme that was in direct violation of the interests of the poor and marginalized they purported to represent.

In terms of the contribution of Canadian NGOs to undermining Aristide's government, Engler and Fenton (2005) have explained this primarily in terms of co-optation by the Canadian state.[16] With most of their funds coming from CIDA, Canadian NGOs have aligned themselves with Canada's political agenda.[17] These activists have also pointed to cultural factors that bind NGO personnel to the Haitian elite, particularly the importance of a common language not spoken by the majority of the Haitian people. There is no question that Canada has acted as an imperialist power in Haiti, and that Canadian NGOs have obfuscated the self-interested actions of the Canadian government. Most adopted an opportunistic double-standard by denouncing Aristide for implementing neoliberal policies while letting Canada off the hook for pushing this approach. But the NGO position also reflected the split in Haitian civil society. Development and Peace did have a well-established network in Haiti and it internalized the perspective of its counterpart NGOs and organizations such as the MPP which opposed Aristide. Development and Peace developed a presence in Haiti in the late 1960s, and Rights and Democracy's work with women's NGOs dates back to the mid-1990s. Canadian NGOs in fact played a very active role in driving Canadian opposition to Aristide through reports, newspaper articles, and parliamentary appearances. They were not simply carrying out the agenda of the state.

The co-optation hypothesis thus has its limits in explaining Canadian NGO behaviour, which seems to have been guided by a combination of principle and opportunism. Ultimately, these NGOs acted recklessly and irresponsibly, much like their counterparts in Haiti. Their willingness to influence a policy of regime change based on flimsy standards of evidence demonstrates the arrogance and ethnocentrism that resides at the heart of democracy promotion, even when it is being conducted by those with a more progressive track record. As we shall see, moreover, all were silent about the campaign of repression waged against Lavalas partisans after the coup had occurred.

---

16   Both authors were active in the Canada Haiti Action Network (CHAN), a group that criticized the policies and democracy promotion efforts of the Canadian government in the wake of the coup.

17   One of the labour organizations involved in the CPH, the *Fédération des travailleurs et travailleuses du Québec* (Federation of Workers of Quebec – FTQ), had direct material interests in Haiti through its investment arm, the *Fond de solidarité* (Solidarity Funds), which owned 12% plus stock options of Gildan, a Canadian clothing company whose main subcontractor were companies controlled by G-184 leader and sweatshop magnate, Andy Apaid (Sprague 2007).

The Switch to Stabilization

State against Popular Civil Society

Once the coup had occurred, the *Gouvernement Intérimaire d'Haïti* (Interim Government of Haiti – IGH) led by Florida-based businessman and de facto Prime Minister, Gérard Latortue, pursued a 'scorched-earth' policy against Lavalas (Dupuy 2005). Several human rights reports documented the IGH's brutality, including eye-witness accounts of attacks by police forces and death squads in the slums of the capital, the remilitarization of Haiti's police, and the freeing of prisoners previously convicted of some of the worst crimes committed by the paramilitaries during the first coup years.[18] Meanwhile, MINUSTAH stood-by and waged its own campaign of repression against the so-called *chimères* of the *bidonvilles*.

The leadership role that Canada and the United States played in reintegrating Haiti back into the international community was central to building the legitimacy of the IGH. This included the articulation of a development framework based around the principle of state failure, which was adopted by the UN agencies, the World Bank and various other donors.[19] It also included high-level diplomatic initiatives to build support for the IGH. According to one high-ranking DFAIT official, Canada convinced other Latin American states to support the UN presence since Haiti was seen as 'a challenge for the region and there was a collective responsibility to act'.[20] But regional organizations such as the OAS and CARICOM and the vast majority of Haitians actually opposed the coup, and Canadian policymakers acted opportunistically to enhance Canada's diplomatic stature and economic interests. Several years into the UN intervention, FOCAL's Executive Director, Carlo Dade (2008), more explicitly drew the link between Canada's leadership role in Haiti and Washington's political agenda: 'the US would welcome Canadian involvement and Canada's taking the lead in Haiti. The administration in Washington has its hands more than full with Afghanistan, Iraq … This is a chance for Canada to step up and provide that sort of focused attention and leadership, and the administration would welcome this.'

---

18   Reports were released by the Institute for Justice and Democracy in Haiti (2004), the Quixote Center (2004), the National Lawyers Guild (2004), the Haiti Accompaniment Project (Flynn, Roth and Fleming 2004), the medical journal the Lancet (Kolbe and Hutson 2006), and the University of Miami School of Law (Griffin 2004). The violence against Lavalas activists and the imprisonment of its leadership even prompted human rights organizations that had been highly critical of Aristide such as Amnesty International (2005) to denounce the turn of events.

19   CIDA (2008a) subsequently developed *An Internal Guide for Effective Development Cooperation in Fragile States* and declared Haiti the only fragile state in the Americas.

20   Interview, Ottawa: 30 October 2008.

Although state building undoubtedly has its merits, building state capacity must be accompanied by efforts to democratize the state, which requires differentiating state institutions from oligarchic or patrimonial control (Shamsie 2008). In Haiti, the opposite occurred – Canada and the United States built the state's security capacity while again assisting oligarchic sectors in imposing their development programme through the *Cadre de Cooperation Intérimaire* (International Cooperation Framework – CCI). Although the CCI was couched in terms of participation by Haitian civil society, it was hastily conceived and overrepresented the G-184 and international NGOs at the expense of majority civil society (Taft-Morales 2005). With a development strategy predicated on increasing foreign direct investment in manufacturing and services and an overall lack of attention to reforming Haiti's agricultural sector, the CCI was an updated version of previous neoliberal development strategies (Shamsie 2009).[21] The programme complemented neoliberal reforms undertaken by the IGH, including a moratorium on income tax for three years and the firing of several thousand public sector employees who had benefited from job creation programmes introduced by Aristide (Hallward 2007).

In terms of US democracy assistance, state-building efforts were complemented by programmes which further developed the capacity of elite parties and civil society organizations in preparation for elections. USAID directly administered approximately $1.5 million in democracy assistance in addition to providing funds to IRI, NDI, and Creative Associates International. These efforts were complemented by a 'cross-cutting' initiative administered by USAID's Office of Transition Initiatives, designed to stabilize the situation through 'quick, visible small projects to promote peace and community cohesion' (USAID 2006a).

The National Endowment for Democracy channelled funds to civil society organizations opposed to Lavalas through two of its partner organizations, the Center for International Private Enterprise (CIPE) and the American Center for International Solidarity (ACILS). The former implemented a project with the ultra-conservative *Centre pour la Libre Entreprise et la Démocratie* (Centre for Free Enterprise and Democracy – CLED) to develop a National Business Agenda, while the latter provided a $100,000 grant to the marginal labour organization, *Batay Ouvriye* (Worker's Struggle) (NED 2005b). During the crisis, Batay Ouvriye had advocated a sort of third way approach, denouncing both the G-184 and the Lavalas government as 'two rotten cheeks in the same torn pants' (Batay Ouvriye 2003). As the organization remained alienated from the larger pro-Lavalas labour movement, the NED likely saw an opportunity to continue exploiting such divisions after the coup had taken place.

With more than $3.6 million in grants from USAID, IRI launched a new political party strengthening programme which channelled funds to a number of

---

21   In June 2005, Canada took the lead in assembling donors at the Montreal Conference to assess the results of the implementation of the neoliberal development framework adopted by the government.

marginal parties. IRI also contributed to a new political discourse whereby Aristide was attributed with occult powers to direct political events from afar: 'On the basis of various reports from Haiti', one IRI report noted, 'the committee is of the view that former President Jean Bertrand Aristide is exerting – through his supporters in Haiti – a negative influence on Haiti's current political process. HIAC [the Haiti International Assessment Committee] is uncertain what the specific evidence is to support this assessment (sic), suggesting a need to probe the exact character of Mr. Aristide's engagement and its purpose' (IRI 2005a).

In 2006, Canada established a Country Development Programming Framework in Haiti which allocated a total of $555 million to reconstruction and development efforts over a five year period (2006–2011), making it the second largest recipient of Canadian ODA after Afghanistan and Canada the second largest donor in Haiti. Within this framework, CIDA also shifted its focus to supporting state institutions, including the *Conseil Électoral Provisoire* (Provisional Electoral Council – CEP), the offices of the president and the prime minister, and the *Office de la Protection du Citoyen* (Ombudsman). In addition, a third fund administered by the Embassy was created, the Democracy and Peace Support Fund with a budget of $5 million from 2006 to 2011, to support civil society organizations and government institutions in the areas of human rights and good governance (CIDA 2009).

CIDA continued to fund several small civil society organizations directly and through various Canadian NGOs. Among the human rights groups that CIDA chose to fund was Haiti's most high profile human rights organization, the National Coalition for Haitian Rights (NCHR), whose reports were directed against Lavalas partisans and whose most serious allegations were subsequently disproved.[22]

CIDA also funded a $325,000 Rights and Democracy project and a second phase of Development and Peace's country programme (CIDA 2006). Through the Rights and Democracy initiative, the organization engaged women's organizations and the anti-Aristide coalition, *Forum Citoyen pour la Réforme de la Justice* (Citizen Forum for Justice Reform), to train civil society organizations on how to develop advocacy plans and public policy proposals. The initiative reinforced the ability of elite civil society to influence the state. Despite its human rights focus, Rights and Democracy failed to denounce the abuses committed by the interim government. Instead, the organization addressed a joint letter with other members of the *Concertation pour Haïti* to Kofi Annan demanding that MINUSTAH restore

---

22 For example, allegations that Aristide's Prime Minister, Yvon Neptune, had orchestrated genocide when supporters of the G-184 clashed with a pro-Lavalas group in the town of St. Marc in February 2004 were subsequently disproven. In its ruling on Neptune v. the State of Haiti, the Inter-American Court of Human Rights criticized both the IGH and Préval government's violations of Neptune's basic human rights, exonerating him of all charges and ordering the government to pay him $95,000 in damages and costs. NCHR's partner office in New York severed ties with its Haitian counterpart for its questionable positions, which changed its name to *Réseau National de Défense des Droits Humains* (National Network for the Defence of Human Rights – RNDDH).

security for elections even as the peacekeeping mission was being criticized for casualties inflicted by its operations in the slums of *Cité Soleil* (Rights and Democracy and the *Concertation pour Haiti* 2006).[23]

Additionally, the Canadian Foundation for the Americas (FOCAL), hosted several conferences in North America on behalf of the Haitian private sector in 2005. With several high-ranking government officials in attendance at one event, including Denis Coderre, Liberal MP and Special Advisor on Haiti to Prime Minister Paul Martin, the conference's conclusions restated old mantras on the role of privatization, enhanced security and the importance of the private sector in reducing poverty (Guevara Rodolfo 2005).

Canada also provided considerable support to the electoral process organized by the interim government. While CIDA's technical and financial support to the CEP and the UNDP to organize the elections were not in themselves objectionable, Canada's silence on the electoral process's shortcomings was. Despite funding an international observation mission, Canada and its partners failed to call international attention to numerous issues throughout the electoral process, including the imprisonment of Lavalas leader Father Gérard Jean-Juste, insufficient voter registration centres, and pro-government forces attacking Lavalas rallies (Concannon 2006).

The results of the 2006 presidential results once again demonstrated the degree of fragmentation of Haiti's political landscape. The *Convergence Démocratique*, which had portrayed itself as a coherent alternative to Aristide, dissolved as an electoral coalition after the coup. Haiti's elite returned to its old pattern of interpersonal rivalry, with 35 presidential candidates appearing on the ballot. Former president René Préval won the election at the head of a new coalition, *Lespwa* (Hope), which included members of the former FL government under Aristide. The majority of FL itself boycotted the election.

During the process, allegations of fraud surfaced when Préval, poised to win the presidency again according to exit polls and the CEP's own initial tabulations – finished 1.3% below the 50% mark needed to prevent a second round of elections. Suspicions were confirmed when television stations broadcast thousands of ballots – mostly for Préval – found in a dump not far from the CEP's tabulation center. Amidst massive protests, a compromise solution apportioned blank ballots among the leading contenders, thereby granting Préval his victory by violating the rules of the game instead of upholding them (Concannon 2006).

---

23   The Deputy Field Commander from September 2005 to September 2006, Eduardo Aldunate (2010), exposes the repressive role that MINUSTAH played after the coup in his recent book, *Backpacks Full of Hope*.

The 'Restoration' of Democracy

The Impasse of Neoliberal Polyarchy

Ultimately, the coup against Aristide and the efforts of the interim government and its international partners to consolidate a functioning neoliberal polyarchy in Haiti failed. True, the campaign of repression against Lavalas helped alter the balance of power between popular and elite social forces by disorganizing the Lavalas base and imprisoning its top leaders. Democracy promotion efforts further strengthened the capacity of elite civil society and parties. But the elite once again failed to achieve a hegemonic position in popular civil society and win the presidency, or, for that matter, even a minimal hegemony within its own ranks to ensure an ideological and organizational unity against the popular masses. In both chambers of parliament, political parties to this day are weak and fragmented and lack party discipline. Those who are elected tend to represent particularistic interests.

Within this simulacra of democracy, neoliberalism remained the dominant paradigm. Indeed, Préval's presidency came to disappoint those who hoped that his victory would mark a rupture with Haiti's long troubled history. The neoliberal approach reasserted through the interim government's CCI was extended through a new IFI-sponsored Growth and Poverty Reduction Strategy Paper supported by Canada, the United States, and Haiti's other major donors (Republic of Haiti 2007). Similar to the CCI, the GPRSP touted the participatory process involved in its creation and emphasized prudent fiscal and monetary policies and exploiting competitive advantages in agriculture, tourism, and textiles. Haiti's loss of food sovereignty was evidenced by the devastating consequences of export-focused, liberalized agriculture, as rapidly increasing food prices led to starvation and food riots in the spring of 2008 (Chinai 2008).

Under Préval, both CIDA and USAID continued to fund elite sectors of civil society largely based in Port-au-Prince. Rights and Democracy initiated a $5.5 million CIDA-funded programme in 2008, which continued to fund its main civil society partners and the *Convention des Partis Politiques Haïtiens* (Convention of Political Parties of Haiti), a coalition of 12 political parties formed in 2005 with a co-opted faction of FL. Rights is also working extensively with youth organizations and partnering with a local NGO to deliver civic education initiatives. At the same time, however, both the United States and Canada further emphasized supporting political institutions. Among other things, a CIDA-funded parliamentary capacity-building programme with a budget of $5 million was implemented. USAID also hired the development consultant, Tetra Tech ARD, to implement a massive decentralization effort to strengthen local government units. Such initiatives have rationalized social contradictions by improving the mechanisms of governance rather than promoting redistributive policies. In the absence of state transformation, they stabilize neoliberalism by creating a more efficient polyarchy.

Indeed, in 2009, the old emphasis on Haiti's low-wage manufacturing sector received renewed impetus with the release of the 'Collier report', a development strategy authored by economist, Paul Collier, Special Adviser on Haiti for UN Secretary-General Ban Ki-moon. The report, also supported by Bill Clinton, now UN Special Envoy for Haiti, noted that Haiti was in a unique position to capitalize on new US legislation providing duty and quota-free access to the American market (Collier 2009). Released shortly before Haiti's parliament began debating a bill to triple the minimum wage to $5 per day (62 cents/hour), the report was seized upon by opponents to keep wages low. Leaked documents indicate that contractors for Fruit of the Loom, Hanes and Levis worked with the US Embassy in Haiti to oppose the increase while the Senate debated the bill, which had already unanimously passed in the Chamber of Deputies. Préval eventually negotiated a two-tier system to pay textile workers a minimum of $3 per day and all other industrial and commercial sectors $5 (Coughlin and Ives 2011a).

Without an organized popular base to sustain his government, Préval himself joined the ranks of the traditional political class. In addition to supporting the international economic programme, he provided unconditional support to MINUSTAH, which increased stability in some of the poorest sectors of Port-au-Prince through a strategy of repression.[24] A leaked US diplomatic cable indicated that the mission also assumed a more direct political role when the head of the mission from June 2006 to August 2007, Edmund Mulet, recommended that criminal charges be filed against Aristide to prevent him from regaining his influence (Mulet was renamed head of mission after the earthquake) (Herz and Ives 2011). The very fact that Haiti's elite required a massive foreign-occupation force to stabilize the country in the absence of civil conflict points to the elite's inability to broker a new social compromise.

Senate elections in April 2009 again illustrated the elite's antipathy towards the popular classes' involvement in Haiti's democracy. When two separate factions of FL submitted a list of candidates, the CEP refused to recognize either one. A joint list was subsequently submitted, but the CEP further required an original signature from the party leader, Aristide, in exile in South Africa. FL challenged the CEP, which does not have the authority to exclude legally recognized political parties, in court. But the CEP ignored FL's legal challenge (and protests from parts of the international community), and held the elections as planned. Although the CEP announced a paltry 11% turnout as many Haitians boycotted the election, virtually every independent observer estimated turnout of less than 5% (Institute for Justice and Democracy in Haiti 2010).

---

24   Although Préval was prepared to align his administration closely with Washington, the impoverished country's efforts to acquire oil at discount prices through Venezuela's Caribbean oil alliance, PetroCaribe, were undermined by the US Embassy under Ambassador Janet Sanderson, who worked with US oil companies controlling distribution networks to sabotage the agreement (Coughlin and Ives 2011b).

Turning a Corner – New Hope for Haiti or Enduring Patterns?

The earthquake that struck Haiti on 12 January 2010 left up to 200,000 dead, 300,000 injured, and 1.5 million displaced. Approximately 20% of federal government employees were killed (Bolton 2011). The international community quickly responded to the earthquake, assuring they would stand as partners with the Haitian government in reconstruction efforts. The United States and Canada pledged, respectively, $2.6 billion and $550 million in assistance, no doubt providing badly needed aid in sectors such as health and education.[25] The Haitian government unquestionably will require international support for some time to come. The question, however, is whether this support will break with previous failed policies, which were at least partially responsible for the state's inability to address the disaster in the first place (Gros 2010).

But politically and economically, Canada, the United States and much of the international community have reverted to the old pattern of bolstering the position of Haiti's transnationalized elite. After the initial shock of the destruction, a ministerial conference chaired by Canadian Prime Minister Stephen Harper was held in Montreal, paving the way for the establishment of the Interim Haiti Recovery Commission (IHRC) and the Haiti Action Plan to guide reconstruction efforts. The plan is in many ways a post-earthquake reiteration of the National Strategy for Growth and Poverty Reduction,[26] with a continued focus on export-processing zones (now called 'Regional Development Centres') as the elite's self-serving panacea to economic development despite its lacklustre results. Although the Haitian state is given a more important role than many previous plans, this is largely limited to establishing enabling infrastructure and implementing an

---

25   Canadian Minister of International Cooperation, Bev Oda, announced a $220 million Haiti Earthquake Relief Fund, which included – confusingly – parts of the $150.15 million and $400 million commitments. The Relief Fund was intended to match the $220 million committed by Canadians for Haiti relief efforts. See CIDA (2010a, 2010b, 2011b). The US government committed $1.1 billion in humanitarian relief assistance and $406 million in recovery assistance to Haiti following the earthquake. This was accompanied by an additional $1.15 billion for reconstruction efforts (USAID 2011b). By September 2010, however, less than 15% of all reconstruction pledges from the international community had arrived ($1.8 billion), of which only 0.3% went to the public sector. The $1.15 billion pledged by the United States also stalled because of partisan politics in Congress (Farmer 2011: 182). Analysing data from the US-government Federal Procurement Data System, the Center for Economic and Policy Research (2011) reported that as of 14 April 2011, 1,490 contracts had been awarded by USAID for a total of $194,458,912. Of those contracts, however, only 23 went to Haitian companies, totalling just $4,841,426, or roughly 2.5% of the total. US Beltway contractors, in contrast, received 40% of the total.

26   Progress has been made on some fronts, however, particularly on issues pertaining to debt. The decision by IFIs to cancel Haiti's (mostly odious) debt can only be seen as positive, though there are concerns that post-quake IMF loans could lead to a new debt cycle (Jubilee USA Network 2010).

adequate incentive policy 'to favour the establishment of manufacturing industries, free zones, industrial areas and areas for the development of tourism' (Republic of Haiti 2010).

One can reasonably argue that in post-earthquake Haiti, any jobs – including those in the export-processing zones – are welcomed. As Shamsie (2010) has argued, however, Haiti has failed to adopt a 'high-road' approach to assembly manufacturing that might contribute to the country's development based on higher wages, unionization, and generous social policies. Among the 23 assembly plants currently in operation in Haiti, only one has a negotiated a collective agreement with its workforce. Without additional protection and state subsidies to increase the standard of living of the workforce, assembly manufacturing will again fail to stimulate the local economy.

In addition to the continuities in development strategy, Canada and the United States both played a role in funding and sanctioning flawed elections. Before the earthquake, the CEP had followed up its controversial decision to exclude FL from the senate election with an announcement that it would exclude 14 political parties – including FL – from the upcoming general election.[27] Although FL followed all registration requirements for this election, the council cited its original failure to submit a party authorization for the April 2009 elections as grounds for its exclusion (Institute for Justice and Democracy in Haiti 2010).

A US diplomatic cable indicates that many of Haiti's international partners were unhappy with the decision. Canadian Ambassador, Gilles Rivard, expressed his concerns at a donor meeting in December prior to the disaster, noting that support for the elections as 'they now stand' would be interpreted by many Haitians as support for Préval and the CEP's decision against Lavalas. 'If this is the kind of partnership we have with the CEP going into the elections', Rivard wondered, 'what kind of transparency can we expect from them as the process unfolds?' (Gurzu 2010).

Despite these protests, the elections were held according to the original decision on November 28, 2010. Only 22.8% of registered voters cast a ballot, indicating, once more, that most Haitians refuse to participate in elections missing the country's most popular political party. Official results after the first round of the presidential election put conservative Mirlande Manigat comfortably in first place, with Jude Celestin from Prévals' INITE coalition in second and the populist-reactionary Michel Martelly in third. After widespread protests in favour of Martelly, an OAS expert mission (six of the seven members of which were from the US, Canada and France) recommended overturning the results based on a review of a sample of ballots.[28] Under pressure from the international community,

---

27   This included a presidential election and elections for all 99 seats in the Chamber of Deputies and 10 of 30 seats in the Senate.

28   Serious methodological issues characterized the analysis conducted by the expert mission. Among other things, the OAS sample did not take into account the results for tally sheets for some 1,326 voting booths which were either never received by the CEP or were

the CEP nudged Celestin out of second place with the highly questionable OAS results, which put Martelly ahead of Celestin by 0.7%. In this context, Martelly emerged victorious in the second round of voting that occurred on 20 March 2011.

Although it is too early to tell what his presidency will mean for Haiti, his well-known ties to the far-right faction of Haiti's elite and recent efforts to re-establish the military – its traditional arm of repression – invite serious concern. The government has also engaged in unlawful evictions of displaced earthquake survivors, including a recent eviction of 514 families taking refuge in a sports stadium (Institute for Justice and Democracy in Haiti and *Bureau des Avocats Internationaux* 2011). Meanwhile, MINUSTAH continues to maintain stability at the expense of meaningful reform. One year after the mission introduced cholera back into the country for the first time since the nineteenth century as a result of scandalous sanitation practices, a contingent of peacekeepers was embroiled in yet another sexual assault scandal (Weisbrot 2011).

Concluding Remarks

Neoliberal polyarchy in Haiti remains fragile, unconsolidated and riddled with contradictions. In the absence of redistribution, democratic political institutions will continue to serve a mostly ritualistic function, legitimizing the power of the dominant class and rotating factions of the political elite. Campaigns of political exclusion and repression will remain the dark side to this sad political theatre.

Haiti thus exposes the extreme contradictions of the model of governance favoured by North American-led regional order in the new millennium. Aligning themselves with Haiti's dominant class, Canada and the United States first sought to destabilize a government which threatened traditional class relations and the neoliberal form of state, before turning to a strategy of stabilization once a more reliable government was back in power. This chapter provided an overview of the tactics of regime change that were deployed during this initial phase, the informal division of labour that occurred between US and Canadian actors, and the shift in strategy that followed. In effect, both countries supported different social sectors in elite civil society in an effort to cement a new hegemonic bloc that might counter the popularity of Lavalas. Although Canadian NGOs acted opportunistically, they had also internalized the perspective of their NGO counterparts in Haiti which opposed the government. Their reckless actions – and the fact that they were never held to account – speak to the democratic deficit at the very core of democracy promotion even when it is pursued by those with more progressive track records.

---

quarantined for irregularities. This corresponded to about 12.7% of the vote. The sample also controlled for irregular tally sheets by excluding them rather than using statistical techniques to estimate voting patterns based on the large amount of electoral data available (see Johnston and Weisbrot 2011).

As democracy assistance programmes moved onto a second phase under the interim government, North American actors continued to support elite civil society groups, many of which legitimized the repressive actions of the interim government against Lavalas partisans. Préval's subsequent acquiescence to international pressure and the interests of the dominant class ensured that transnational capital continued to dictate the economic programme while Lavalas remained excluded from political life. Stabilization tactics began to reinforce the legitimacy of the state before Haiti was hit by the terrible earthquake in January 2010. In its wake, the United States and Canada have continued to push agro-exports and assembly manufacturing despite the lacklustre results of this programme.

Haiti's extreme subordinate position in the global capitalist economy and the regional order of the Americas undoubtedly made possible the overtly political approach of Canada and the United States. The inability of the elite to lead a hegemonic process under its own leadership provided the political context in which aggressive forms of interventionism were required to save the neoliberal state. The situation in Peru, however, has been very different.

Chapter 4
Building Inclusive Neoliberalism in Peru

As left and centre-left governments came to power across the Americas in the early 2000s, Peru remained a staunch North American ally and a bulwark against left 'populism'. In the wake of Peru's transition back to democracy in 2000, presidents Alejandro Toledo and Alan García deepened the process of neoliberal transformation initiated by the authoritarian regime of Alberto Fujimori, and spoke out widely against their leftist neighbours in Bolivia and Venezuela. Neoliberal policies have not been without success in Peru – since 2000, the country's economic performance has been among the most impressive in Latin America. In the two years prior to the global recession of 2008, annual GDP growth stood at 8% (Economist Intelligence Unit 2007). Behind the commodity-driven expansion, however, lies the reality of growing political instability, social conflict and high-levels of poverty. Across Peru's diverse geographical landscape in the mountains of the Andes and the jungle of the Amazon, popular social forces have mobilized against the state and mining companies after a decade of repression and disorganization suffered during the Fujimori years.

This chapter critically examines democracy promotion and assistance in a country where the internal conditions for the consolidation of neoliberal polyarchy have been more propitious. It expands the neo-Gramscian critique by historically analysing North American democracy assistance programmes in a country where the hegemonic aspirations of the elite have not encountered the same level of resistance as in neighbouring states, and where the democracy wars have been waged through a policy of containment rather than regime change.

In this political context, US and Canadian democracy assistance over the last decade have reinforced the state's efforts to build a new hegemonic order through a strategy of stabilization. As such, they have complemented the foreign policies of both countries, which have focused on deepening the neoliberal model through FTAs and mining expansion. For the United States, Peru's co-operation in the regional War on Drugs has also been an important consideration behind its efforts in building the legitimacy of the state.

This chapter will explore the kinds of tactics democracy assistance programmes have deployed to contribute to elite hegemony, and will theorize the more subtle ways in which North American democracy promotion has sought to stabilize neoliberal polyarchy. It will pay particular attention to the ideological nature of the hegemonic project with which democracy assistance programmes have been associated, and will consider how US and Canadian approaches varied. The chapter is divided into three parts. The first part provides a brief overview of the state and social relations that emerged following the democratic transition in 2000, and Peru's bilateral relations with Canada and the United States. The second part then examines the ways in which soft tactics of democracy promotion contributed to the production of hegemony during the Toledo presidency (2001–2006). I argue that, from the outset, both Canada and the United States contributed to the state's project of passive revolution by lending support to the government in creating new forms of inclusive neoliberal governance while rationalizing and strengthening traditional liberal-democratic institutions. Both also supported civil society organizations rooted in the urban professional and middle class sectors, some of which were uncritical of neoliberalism but others which were led by left intellectuals with more activist backgrounds. Although they have promoted important reforms, even the more progressive ones have contributed to the state's project of passive revolution by socializing subaltern groups into liberal citizenship norms that place safe limits on structural transformation (though some have also been openly critical of García's authoritarian backsliding).

During the Toledo years, there were also important differences between US and Canadian approaches. Some Canadian NGOs supported civil society organizations more critical of the state, and continued to do so after García had come to power. Development and Peace in particular supported radical organizations with a significant social base that mobilized against the neoliberal model. The United States also focused on building the legitimacy of the Toledo government directly, whereas Canada's approach to stabilization focused more on building state institutions, including in the mining sector. Recent developments indicate, however, that the Harper government is now mobilizing more compliant Canadian NGOs in Peru to work with Canadian mining companies as part of its overall reconfiguration of the Canadian development field of practice.

The third section explores how the United States reoriented its democracy promotion efforts to contain both the spread of 'anti-systemic' forces in Peru and left 'populism' in the region at large after left-nationalist Ollanta Humala's near presidential victory in 2006. During the García presidency (2006–2011), a new focus on rebuilding political parties and incorporating Humala's *Partido Nacionalista Peruano* (Peruvian Nationalist Party – PNP) into the mainstream emerged, as well as a new framing of democracy around transparency and accountability in civil society. These efforts failed to reduce Humala's popularity, however, and many were hopeful that his victory in the 2011 presidential election would mark the beginning of a more inclusive order. I end with some comments on why it appears the United States had very little to fear all along.

State and Civil Society Following the Transition (back) to Democracy

Peru is the most geographically diverse of the three countries examined closely in this book. The prosperous urban settlements along the Pacific coast are hemmed in by the hulking Andean mountain chain to the east, on the other side of which lies the vast *selva* (jungle) of the Amazonian rainforest. While the vast majority of Peru's 27.9 million inhabitants speak Spanish, in the Andean region Aymara and Quechua continue to be spoken widely in addition to the numerous languages spoken by the indigenous peoples of the Amazon. Nearly half the population is of indigenous descent, about 15% European, and 37% mestizo (Central Intelligence Agency 2012). Although most of the population is engaged in subsistence farming, Peru is extremely rich in mineral deposits and has a fairly-well developed financial and commercial sector. Its natural resource base includes natural gas, hydrocarbons, zinc, iron, lead, gold, timber, and silver; Peru is also one of the world's largest copper suppliers. But its popular classes have not shared the benefits reaped through its natural wealth exports, which have enjoyed particularly favourable prices in international commodity markets throughout most of the last decade. In 2009, the poorest 20% of Peruvians held only 3.9% of the country's income share while the richest 10% held 35.9% (World Bank 2012).[1] The class divisions have taken on a striking geography in the capital city, Lima, where the country's wealthy (primarily white) financial and commercial oligarchy dominate the lush suburbs and the poorest are confined to the desert shantytowns that encircle the city.

Both the United States and Canada have significant investment in Peru and growing trade relations. In 2008, the United States exported $6.2 billion while importing $5.8 billion in return. This represented a significant increase in trade relations from 2000, when US exports equalled $1.7 billion and imports were $2 billion (US Census Bureau 2010b). US direct investment in Peru reached $3.9 billion in 2008, making it the second largest source of foreign direct investment after Spain (US Department of State 2008a). Major US mining companies include Newmont Mining Corporation and Phelps Dodge Corporation. Peru is currently Canada's third largest trade partner in Latin America. In 2008, Canadian exports to Peru equalled $382.5 million while imports totalled $2.5 billion. However, the total stock of Canadian direct investment in the same year was estimated at $1.8 billion, most of it concentrated in extractive industries. Canadian mining companies are amongst the most dominant in the Peruvian mining sector, leading explorations and investment in gold-mining operations (Government of Canada 2009b). The Canadian-based Barrick Gold Corporation – the largest gold mining company in the world – runs major operations throughout Peru.

Since the transition back to democracy in 2000, Peru has been governed by presidents who have remained committed to a neoliberal development model, rendering it somewhat of an anomaly in a region that has largely shifted to the left. Over the past several years, the country has been marked by widespread, growing

---

1 The García government claimed to have reduced poverty from 48.7% to 34.8% between 2005 and 2009 based on a report produced by Peru's National Institute of Statistics and Informatics. The former head of the institute, however, rejected the statistics as artificial, pointing to figures on food poverty that he argued were more reliable. Based on these figures, poverty increased from 28.6% to 32.9% overall and from 40.7% to 45.8% in rural areas (Skeen 2010a).

political and social instability rooted in the forms of exploitation and domination that continue to define the relations between social classes, ethnic groups, the state and society.

In the 2001 presidential election, Alejandro Toledo defeated former president Alan García on the ticket of Peru's oldest party, the *Alianza Popular Revolucionaria Americana* (American Popular Revolutionary Alliance – APRA).[2] With national peasant and labour movements disorganized after a decade of authoritarian repression during the internal war against the communist-terrorist organization, *Sendero Luminoso* (Shining Path) (Yashar 2005), Toledo's party, *Perú Posible* (PP), offered modest social programmes and an increase in social expenditures as a percentage of GDP in a neo-populist appeal to the poor (Barr 2004). Yet Toledo's economic plan played to the interests of foreign capital and Peru's predominately white, coastal-based commercial, financial and agro-mineral oligarchy, which supported the President's strong free trade agenda and efforts at privatization. Negotiations for the Peru-US Trade Promotion Agreement (PTPA) also began in 2004 and the agreement was approved by Peru's Congress in the final month of Toledo's presidency in 2006 (the US Congress approved the agreement in 2007).

By his last year in office, Toledo's approval ratings fell to the single digits. As Weisbrot (2006) reported on the eve of the 2006 presidential elections, GDP per capita was about the same as it was in 1981 and the poverty rate had only marginally decreased. In Lima and Callao, poverty had actually increased from 31.8% to 36.6%. Social protests, meanwhile, climbed from approximately 250 in 1999, to 680 in 2000, and 810 in 2002 (Arce 2008: 42). According to the office of the human rights ombudsman, the *Defensoría del Pueblo* (2007), the majority of mobilizations (46%) have been against foreign direct investment, with communities protesting the social and environmental consequences of specific mining operations.

In the 2006 elections, García again ran for the presidency, this time narrowly defeating left nationalist, Ollanta Humala, in a second-round vote, thus marking the return to power of Peru's main political party in 2006 after sixteen years (Cameron 2008).[3] While his campaign still paid tribute to elements of APRA's traditional left discourse, he quickly moved to consolidate the alliance with big business. Among other things, business leaders and technocrats from the Toledo

---

2   Toledo's rise from humble indigenous shoe-shine boy in the city of Chimbote on Peru's northern coast to PhD at Stanford University and international civil servant initially provided him with considerable popular appeal (Barr 2003).

3   Humala is a former Lieutenant Colonel who fought in the internal conflict against the Shining Path in 1992. In October 2000, he led an unsuccessful military uprising against Fujimori for which he was subsequently pardoned by Congress. Both his father and brother, who led a local uprising against the Toledo government in 2005, are prominent figures in the small indigenous-nationalist movement, the *Movimiento Nacionalista Peruano*. In the second round of elections in June 2011, Humala defeated Keiko Fujimori, daughter of the former autocrat now serving a 25-year sentence for human rights abuses.

government were kept in key government positions and a new public sector austerity package was introduced. García's economic plan focused on commodity exports, which benefitted from high mineral prices in the international market. In 2009, an FTA with Canada also came into effect.

Despite its rapidly expanding economy, however, Peru continued to be a major recipient of aid from both Canada and the United States.[4] In 2007–2008, it received $53.5 million from USAID (Department of Commerce 2009) and $20 million from Canada in bilateral and multilateral development assistance (Government of Canada 2009b). Although Fujimori had been a key regional ally on free market reforms, security issues, and the War on Drugs, the administration of George W. Bush provided millions in support to help guide Peru's transition back to democracy. Subsequently, both the Bush administration and the Obama administration continued to spend large sums in democracy assistance; from 2001 to 2008, USAID spent over $60 million in democracy assistance and an additional $20 million from 2009–2012. Under Liberal and Conservative governments, Canada spent a total of $38.3 million from 2001 to 2009.[5]

García's government took a repressive turn in response to protests against the neoliberal development model. In addition to a general criminalization of dissent (Laplante 2008), the government reacted with fierce repression to indigenous mobilizations in the summer of 2009 against a series of legislative decrees relative to the implementation of the PTPA, which, among other things, made it possible to sell 64% of the forests of Peru to transnational corporations. Indigenous communities in the Amazonas responded with two months of protests, culminating in clashes with police and military forces and the deaths of more than 20 protestors in the so-called Bagua massacre of June 2009 (Zibechi 2009). As the events unfolded, both the United States and Canada remained silent.

Democracy Assistance During the Toledo Years

The United States Seeks to Legitimize the State

USAID's first programme in the wake of the transition from Fujimori's authoritarian regime was implemented by the Office for Transition Initiatives (OTI). According to statements made by a political officer of the US Embassy: 'US objectives in the post-Fujimori period were to stabilize and facilitate the democratic transition and to recover the US image.' The final report of the USAID programme noted that

---

4   It remains one of 20 countries of focus for Canadian development assistance and is a key partner in the context of the Harper government's Americas Strategy.

5   US figures are drawn from USAID congressional reports (2002a, 2003, 2004, 2005b, 2006a, 2007a, 2008b, 2009, 2010, 2011a). The calculation is based on the budgetary estimate for 2009 and the USAID request for 2012. Canadian figures were provided to me by the CIDA Directorate for the Americas (6 January 2010).

the 'OTI was very useful politically to the Embassy because it had quick funds available for projects for public diplomacy and for projects of interest to senior Peruvian government officials' (Hill, McBride and Albertini 2003). In addition to the programmes directly supporting the government which will be examined below, US democracy assistance was initially focused on human rights issues, including support to the human rights' ombudsman and the truth and reconciliation commission (USAID 2002b).

US support to decentralization, which was launched in November 2002 and transformed the country's departments into 24 regions with elected presidents, was particularly important. The decision-making prerogatives of regional, provincial, and district councils were expanded and new mechanisms for civil society participation at each level of governance were created (Arce 2008). The Toledo government's support for decentralization seems to have at least partially been constructed based on what the Peruvian scholar, Isabel Remy (2005), describes as a strategy of redirecting social policy demands away from the central state to the regional and local levels to manage social tensions. Decentralization was accompanied by a modification of the law surrounding control over royalties generated from mining and hydrocarbon companies, the so-called *canon*. In the new arrangement, the bulk of the *canon* revenues are allocated to the municipalities and regions in which they are generated (Poder Legislativo 2004). This has led to a breakdown in solidarity between regions as new 'horizontal' tensions have emerged, in addition to 'vertical' class tensions that have been relocated to the local territorial level (Remy 2008). The focus on the local also occurs at the expense of any meaningful participation in national decision making.

US support to decentralization began with the OTI programme and was significantly enhanced through the $20 million Pro-Decentralization Project (PRODES) implemented by ARD Inc. The overall approach of the programme was technocratic, focusing primarily on improving the administrative efficiency of different units of government through technical assistance and training and managerial workshops (USAID 2008c). USAID's role in the decentralization process seems to have largely been guided by the same instrumentalist preoccupations that led the Toledo government to launch the process in the first place – the high level of social conflict and the need to create new mechanisms of local inclusion in the context of regional disturbances which challenged the authority of the central state. One USAID official who was interviewed acknowledged that the agency's support to decentralization has been guided by a preoccupation with reducing social tensions and that the policy has faced difficulties in part because the issue of redistribution has not been adequately addressed by the decentralization scheme.[6]

The decentralization project can also be a read as an attempt to produce the hegemony of inclusive neoliberalism at different geographic scales. Legoas' (2007) analysis of the discourses associated with the decentralization process calls our attention to the type of citizen being constructed through this initiative.

---

6   Interview, Lima, 8 January 2009.

For example, one of the main Peruvian organizations to be involved in USAID's programme was the *Grupo Propuesta Ciudadana* (Group for a Citizen's Proposal – GPC), which promoted a 'watch dog' conception of citizen participation through popular education programmes based primarily on monitoring authorities to prevent corruption and inefficient use of resources.

US democracy assistance programmes in the area of democratic governance also articulated in large part around re-establishing the credibility of the legislature. Through the OTI programme, Congress received a total of $387,619 in assistance (the executive, for its part, received $197,899) (Hill, McBride and Diaz-Albertini 2003). Such efforts have been framed specifically in terms of the inability of the political class to lead a process of congressional reform of its own volition (CONSODE 2003). To trigger this process, USAID played a key role in organizing and funding a coalition of organizations, the *Consorcio Sociedad Democrática* (Consortium for a Democratic Society – CONSODE), which included five Peruvian NGOs as well as NDI, funded by the NED, and lasted from 2001 to 2006. CONSODE sought to enhance the link between congressional representatives and the citizenry in order to increase popular support for the legislature and improve its image through workshops and publishing guides designed to socialize *congresistas* on the role of legislators in a representative democracy (CONSODE 2004). USAID also funded the Center for International Development (2003) of the State University of New York to implement a programme, Developing Skills of the Peruvian Congress, which contributed to the development of a public relations strategy for a congressional office charged with this responsibility.

A particularly glaring example of USAID's attempt to manipulate public opinion in favour of the government was IRI's Promoting Political Stability by Improving Government Communications. According to one programme document: 'Although Peru has experienced notable levels of economic growth in the past two years, the Toledo administration has not been effective in communicating the progress achieved. To address this issue, the United States Agency for International Development (USAID) authorized IRI to deploy a specialized task force of political communications experts' (IRI 2005). One IRI board member, Richard S. Williamson, testified before the Committee on International Relations of the House of Representatives that the programme was implemented specifically in reaction to the mass mobilizations in Bolivia that had led to the toppling of the government (House of Representatives 2004).

The focus on bolstering the legitimacy of the government through democracy assistance also overlapped with the US' anti-drug strategy. In 2002, Peru signed the Andean Trade Promotion and Drug-Eradication Act, which renewed and modified the Andean Trade Preference Act in effect since 1991. Under the new agreement, an aggressive eradication strategy pushed by the United States won out over the Peruvian government's proposal to focus more on development efforts to provide impoverished *cocaleros* (coca farmers) with a viable alternative. The forced eradication plan launched in January 2003, however, led to a backlash on the part of *cocaleros* increasingly organized at the regional and national level.

Fearing a political crisis similar to the one afflicting neighbouring Bolivia, Toledo backed down and announced a new plan to carry out eradication in a 'concerted, gradual' fashion with incentives for farmers to carry out eradication (Obando 2006). Meanwhile, the United States supported efforts on the part of the Peruvian government to conduct a campaign 'to increase public awareness of the role coca farmers play in drug trafficking, and of negative impacts of drug trafficking on Peru'. A report boasted that 'this awareness has contributed to weakened support for political organizations of coca growers ... who have carried out violent resistance to eradication' (Taft-Morales 2009).

Framing the Public Discourse in Civil Society

In civil society, the effort to stabilize Peru's neoliberal polyarchy was bolstered by supporting NGOs and coalitions that contributed indirectly to the state's project of passive revolution. Prior to the transition, the United States had supported civil society organizations such as the *Coordinadora Nacional de Derechos Humanos* (National Coordinator for Human Rights – CNDDHH) and many of its members, a coalition of over 60 organizations founded in 1985 which served as the main protagonist in Peru's struggle for a return to democracy throughout the 1990s. According to Drzewieniecki's study of the organization, its member organizations, led mostly by university-educated middle and upper class activists, increasingly narrowed the human rights discourse to civil and political rights by the late 1980s (Drzewieniecki 2002). This emphasis likely reflected a confluence of factors, including the elite-orientations of the organizations themselves, a general process of professionalization on the part of those with more radical tendencies, and a natural reaction to the lived experience of state repression.

US support to human rights organizations helped to cement a new moral-ethical order based on restrictive albeit important elements of democratic governance. As Sum 2008 emphasizes, the production of hegemony in a general sense is a process that involves co-constitution and co-evolution of subjects and objects insofar as popular concerns and visions are selected, retained and reinforced that do not conflict with underlying material interests. This was certainly the case in Peru, where those actors with more radical conceptions of democracy began to narrow their democratic discourses. The case of two human rights NGOs with a more radical orientation – the *Instituto Peruano de Educación en Derechos Humanos y la Paz* (Peruvian Institute for Education in Human Rights and Peace – IPEDEHP) and *Movimiento Manuela Ramos* (Manuela Ramos Movement – MMR) illustrates how even those which have traditionally been on the left have pragmatically aligned with donor concerns.

As one of the most prominent and politically radical organizations of the CNDDHH, IPEDEHP was established primarily by progressive educators on the left with the goal of fostering democratic values and respect for human rights throughout the country. Over the years, it has received a steady stream of US funding through various initiatives. According to an IPEDEHP founder, US support to human rights organizations following the transition represented a 'convergence of interests' between internal human rights concerns and the vision of democracy of the United States and other donors. But on economic, social and cultural rights there were often tensions, with many feeling that the US was exacerbating rights violations through its economic policies.[7] Yet IPEDEHP and others who have received US funding tend not to publicly criticize US economic policies. To the extent that social and economic rights are discussed by these NGOs, they are framed in a more general sense as unfulfilled obligations on the part of the state. They are abstracted, in other words, from the internal and external social forces that impede their realization, with the exception, in some cases, of transnational mining practices. Rather than mobilize oppressed communities politically, they implement workshops that seek constructive solutions with the state while creating new democratic liberal subjectivities.

The case of MMR – a feminist organization established in 1978 as one of the main organizations to emerge from the women's movement – illustrates more specifically the trend towards professionalization and the pressures of conformity induced by donor priorities (Barrig 2002). MMR has received funding from USAID since the 1990s; it received a grant from the OTI in the early 2000s of $85,812 for its work in strengthening democracy and has continued to receive USAID funding since. Led by middle-class women, MMR's leaders were at one point associated with the left and have remained committed to empowering poor and marginalized women, particularly in the areas of health and reproductive rights.

According to a staff-member of MMR, the organization has had a transformative impact on the public discourse on gender relations, making violence against women unacceptable – even if abuses still occur behind closed doors.[8] While this may be true, MMR has equally contributed to a restrictive notion of economic rights that fits perfectly well within the ideology of its main donor. Its focus on economic rights has been framed entirely in terms of supporting micro-financing initiatives, which arguably has the effect of calling attention away from class relations sustaining inequality and discursively re-constructs poor women as mini-entrepreneurs who simply require access to small amounts of capital to more effectively participate in the market. Seen from this perspective, MMR has helped disseminate a gendered ideology of liberal capitalist citizenship.[9]

In addition to producing a new hegemony based on a human-rights vision of democracy, US programmes also reinforced the hegemony of the private sector. USAID's support to the national business association, the *Confederación Nacional de Instituciones Empresariales Privadas* (National Confederation of Private

---

7   Interview, Lima, 13 January 2009.

8   Interview, Lima: 16 January 2009.

9   The organisation has also been willing to modify its own commitment to reproductive rights, as demonstrated by its decision to drop its public advocacy of abortion rights as a result of the US government's 'global gag rule' under the Bush administration (see Mollman 2003).

Business Associations – CONFIEP) – a key ally of the Fujimori government until the very end – dates back to its establishment in 1984 (Morón and Sanborn 2006). Through the OTI programme, USAID provided a grant of $287,000 to the Lima Chamber of Commerce to implement a legal aid programme (Management Systems International 2000). Both the National Endowment for Democracy (NED) and its sister organization, the Center for International Private Enterprise (CIPE), have also provided grants to business associations dating back to the mid-1980s, including pro-business think tanks such as the *Instituto Libertad y Democracia* (Institute for Liberty and Democracy – ILD), founded by Peru's most famous neoliberal intellectual, Hernando de Soto. In the early 2000s, CIPE began providing grants to other private sector organizations, including CONFIEP itself. From 2002 to 2007, a total of $5.5 million was channelled by CIPE to pro-business groups, a transfer of funds which Minella (2009) argues contributed considerably to the hegemony of the private sector in economic policy debates.

The Canadian Approach to Stabilization

CIDA has also focused its democracy assistance on supporting public institutions and good governance. Because Canada's programmes – launched by Liberal governments but largely extended by the Harper government – did not change from the Toledo to the García years, we will look at them as a whole before returning to the shifting US approach. Like USAID, CIDA supported the Truth and Reconciliation Commission when it was first put into place and has remained an important funder of the human rights ombudsman, the *Defensoría del Pueblo*, which has played an important role in monitoring and responding to human rights abuses.

Yet, tensions between Canada's support to human rights instruments and the economic objectives of its multinationals have been internalized in its democratic governance programming framework. During the Toledo years, CIDA began contributing to a process of redefining the regulatory framework governing hydrocarbon and mineral extraction through several governance programmes. From 2003 to 2008, CIDA provided an $8.75 million grant to the Canadian Petroleum Institute (CPI), a non-profit organization that provides management and technical training to international petroleum companies. Among other things, the project implemented by the CPI was intended to promote potential oilfields to attract new investment by petroleum companies, including Canadian companies, which, according to one CIDA statement, were 'already participating in the project and gaining a competitive edge in the process' (CIDA 2006).

Likewise, a $13.6 million project implemented by the Canadian consortium of Roche Limited, Golder Associates Limited and the Association of Community Colleges of Canada from 2003 to 2011 provided technical assistance to Peru's Ministry of Energy and Mines on policy and regulatory reform issues related to the minerals and metals sector. The stated aim of the Peru-Canada Mineral Resources Reform Project (PERCAN) was to improve the sustainable development of the mining sector through mechanisms to increase dialogue between local communities, municipal governments and mining companies and voluntary corporate social responsibility (Government of Canada 2009c) (Ministerio de Energia y Minas and CIDA). Among other things, the project led to the creation of 'toolkits' designed 'to help communities and public institutions get a better grasp of the risks and opportunities of extractive industries and, in so doing, help mitigate social conflict' (Government of Canada 2009c).[10]

The regulatory framework put in place by the Peruvian state, however, has proved wholly in adequate at mitigating exploitative practices in the mining sector (Salazar 2008). For its own multinationals, Ottawa has consistently favoured promoting voluntary standards of conduct in the framework of corporate social responsibility rather than subjecting them to enforceable legal norms (as discussed in Chapter 2). This has contributed to a new strategy of governance in which a liberalized regulatory structure coexists with loose environmental, health and safety standards.[11] This has also been the preferred approach of the United States, which situates its support to the mining sector in Peru within the international voluntary Extractive Industries Transparency Initiative (EITI) framework. Canadian companies themselves have consistently taken advantage of lax regulations and are regularly criticized by Canadian NGOs for their involvement in mining operations that have violated community rights in both Peru and Bolivia (Mining Watch Canada 2005, North, Patroni and Clark 2006b). Among the many clashes that have occurred between local communities, workers and Canadian mining companies in Latin America, Gordon (2010) cites two conflicts with Barrick Gold in Peru. In May 2006, community members in Huallapampa blocked road access to Barrick's mines when it refused to grant a pay raise; police forces reacted with fierce repression, killing two protestors in the process. Barrick's activities have also been the target of protests in the district of Quiruvilca in the department of La Libertad for its failure to produce promised jobs and for contamination of clear water supplies (Arnold 2012). In April 2007, two more protestors were killed when thousands marched in the Ancash region to demand the cancellation of contracts with Barrick and another company, Antima. The Peruvian NGO, *CooperAcción*, has been particularly critical of Canadian mining activities. In 2008, it argued that, in the region of Amazonas where new concessions were rapidly expanding, exploration conducted primarily by Canadian companies would most certainly lead to increased social conflict (Salazar 2008).

---

10 Unfortunately, additional details on the initiative are hidden from the public. An Access to Information Request for a project report was denied. No rationale was provided, but two clauses of the Access to Information Act were cryptically invoked: 13(1) a government of a foreign state; and 15(1) international affairs and defence. It is reasonable to infer that CIDA is aware that the project approach may be controversial, and that it prefers not to open itself up to criticism.

11 See, for example, Canada's corporate social responsibility strategy for the Canadian international extractive sector (DFAIT 2009).

In the aftermath of the government's repressive response to the uprising in the Amazonas in June 2009, three Canadian NGOs[12] launched a campaign to protest the Canadian government's intent to move ahead with the ratification of its own FTA with Peru despite the opposition to the agreement with the United States. They observed that the legislative decrees will benefit Canadian mining companies enormously since half of Amazonian territories leased for petroleum exploration and extraction have gone to Canadian companies. The Alberta-based petrochemical firm Petrolifera is one such beneficiary, having signed an agreement with the Peruvian government recently to explore land inhabited by one of the world's last un-contacted tribes. Despite concerns in both Canada and Peru, the FTA was ratified, with Canada staying silent on the massacre in the Amazon (The Council of Canadians 2009). Nonetheless, the indigenous protestors did manage to force the Peruvian Congress to respond to their demands by repealing some of the decrees (Carlsen 2009c).

In advancing the interests of Canadian-based multinationals in Peru, Canada has involved itself in longstanding struggles between political and economic elites closely aligned with the state against indigenous movements that have vigorously denounced exploitative mining practices. At the programming level, the term governance has provided a sufficiently elastic concept to link the management of key resources – an issue that should be determined according to internal policy debates – with the very notion of democracy. Uninterested in creating a truly inclusive order, Canada has helped build the institutional capacity of the Peruvian state to manage social conflict by creating new mechanisms through which neoliberal hegemony is articulated.

But there are differences in the US and Canadian approaches. Canada's efforts have focused less on building the legitimacy of the government and more on strengthening the institutions of the state and mediating the relationship with civil society. Like their American counterparts, CIDA development officials have expressed serious concerns with the lack of social stability in Peru. Yet, they do not view their activities as being political. One CIDA official in Lima went so far as to say that she was wary of the US approach to supporting democracy, which she argued could easily be perceived as partisan given its strong focus on rebuilding political parties, and that US democracy agencies are themselves often perceived as extensions of the CIA.[13]

At the same time, the differences may be more of degree rather than kind. The Canadian and US ambassadors in Peru have worked together, for instance, to coordinate the efforts of mining companies and embassies with mining interests to counter threats from 'radical forces'. According to a leaked cable, the ambassadors

---

12   The NGOs were the Council of Canadians, Common Frontiers, and Mining Watch Canada.

13   Interview, Lima: 22 January 2009. CIDA actually does indirectly engage in support to political parties through International IDEA, to which it contributed $97,239 for 2005–2008 (CIDA 2009).

co-hosted a meeting on 11 August 2005, for representatives of international mining companies, where they lamented the radicalism of local and international NGOs, religious leaders, and the Communist Party. The Canadian Ambassador, Geneviève des Rivières, noted that many NGOs such as Oxfam seemed to be ahead of the mining companies in consulting polling institutions to obtain a feel for what issues and concerns motivate communities, and how and what messages to convey in rural mining regions. The ambassadors encouraged mining companies to compile their own lists of activists engaged in 'violence', and expressed their pleasure surrounding the appointment of a new prime minister, Pablo Kuczynski, who was prepared to tackle the lawlessness (*The Guardian* 2011). For both the United States and Canada, the structural outcome of their actions is arguably the same – the de-politicization of social conflicts and the stabilization of the neoliberal state. With Canadian NGOs now partnering with Canadian mining companies, this process is intensifying (discussed below).

Supporting both Elite and Popular Civil Society

Canadian support to civil society organizations has largely targeted the same or similar types of NGOs as US democracy assistance, albeit with some important distinctions.[14] First, Canada has not situated its support to civil society within a clearly-defined diagnostic of the barriers to democracy and the need to enhance political stability. This has meant that Canadian support to civil society organizations has been less focused on specific areas of democracy, encompassing NGOs working on diverse thematic issues rather than coalitions advancing democratic reform. Some of these organizations advance a more expansive notion of citizenship than organizations funded by the United States. Second, Canadian civil society support has not been channelled to private-sector associations. Finally, Canadian development assistance through the NGO, Development and Peace, has empowered grassroots organizations associated with the Catholic Church that have a larger social constituency than most NGOs and which have articulated a cohesive economic alternative to neoliberalism.

One of CIDA's most important civil society partners has been the *Consorcio de Investigación Económica y Social* (Economic and Social Research Consortium – CIES), an umbrella organization that brings together over 30 social and economic research institutions in Peru to enhance the level of national debate on key economic and social policy alternatives. Through the Ottawa-based International Development Research Centre (IDRC), CIDA provided the IDRC-CIES with a $4.9 million grant from 2004–2009. For the most part, the consortium adopts a mildly critical approach to public-policy issues emphasizing themes such as poverty reduction that fit within

---

14 NGOs which have received both Canadian and US support include IPEDEHP, which received a grant of close to a $100,000 for 2005–2007, and MMR, which received a grant of $96,000 for 2005–2006 (CIDA 2009).

the agendas of the state and international donors.[15] IDRC has also supported a research project carried out by a small group of researchers in each of the Andean countries entitled *Gobernabilidad Democrática en el Región Andina* focusing on social movements. The reports published by the project are politically diverse, with many criticizing neoliberalism and supporting radical governments in the region and others vociferously denouncing leaders such as Chávez.[16]

Gender equality and environmental issues have also been a major focus of CIDA programming.[17] In addition to its support to MMR, CIDA (2009) provided a grant for $96,000 to the *Centro de la Mujer Peruana Flora Tristán* (Centre of the Peruvian Woman Flora Tristán), an NGO founded in 1979 and named after the nineteenth-century Peruvian feminist socialist. *Flora Tristán* emerged from the same social sector of middle-class urban feminist activists associated with the left as its sister organization, MMR.[18] Among its many important activities, *Flora Tristán* promotes the full-spectrum of women's rights, with a strong focus on reproductive and sexual rights, as well as gay and lesbian rights. *Flora Tristán* has helped introduce new concepts into the public discourse, including the notion of 'feminicide' to describe the murder of women in conditions of gender-based discrimination and violence. It has also linked this concept to neoliberalism and the commodification of women within the logic of the market (Meléndez López and Sarmiento Rissi 2008), though it does not criticize Canada or the United States specifically for their support of this model. Like other progressive NGOs, however, *Flora Tristán* has not effectively linked its social concerns with an actual strategy of popular mobilization, focusing more on capacity-building initiatives that fit within the developmental logic of NGO projects.

Canada has also supported the *Centro de Estudios para el Desarrollo y la Participación* (Centre of Studies for Development and Participation – CEDEP), a Lima-based NGO which has received grants from numerous bilateral and international donors, including USAID. Canada provided the NGO with a grant of nearly $100,000 for 2005–2007 for a project promoting citizen participation in decentralized governance and gender equality (CIDA 2009). CEDEP has taken a radical position on public issues through its publication, *Socialismo y Participación*, which regularly criticizes neoliberal policies and transnational

---

15   This claim is made based on a sample review of CIES's main publication, *Economía y Sociedad*, published on a trimester basis. The review is available on the CIES web-site at: http://www.cies.org.pe/publicaciones/economia-y-sociedad?page=1.

16   Readers can consult the project's virtual library: http://www.gobernabilidadandina. org/biblioteca.php.

17   In the area of human rights, CIDA has also provided support to the *Instituto de Democracia y Derechos Humanos de la Pontificia Universidad Católica del Perú* (Democracy and Human Rights Institute of the Catholic Pontificate University of Peru – IDEHPUCP). For 2005–2007, IDEHPUCP was given a grant of $97,351 (CIDA 2009).

18   Interview with representative, Lima: 21 January 2009.

corporations. The publication has also offered a more nuanced perspective on Ollanta Humala (see, for example, Reyes 2008), who was largely vilified in the mainstream media prior to his presidential victory. Ultimately, however, CEDEP's activities at the grassroots level focus mainly on capacity-building exercises rather than social mobilization.

In the area of the environment, CIDA has supported *Asociación Civil Labor* (Civil Labour Association – ACL), an NGO with a fairly comprehensive view of the linkages between economic and social justice and sustainable development that began as a labour collective before becoming an NGO in the 1990s. For 2005–2007, the organization was given nearly $100,000 by CIDA (2009) to implement a project intended to strengthen relationships between local communities, local government, and mining enterprises. Canada has also supported the environmental NGO, *Sociedad Peruana de Derecho Ambiental* (Peruvian Society for Environmental Rights – SPDA), which obtained nearly $100,000 for 2005–2007 (CIDA 2009). SPDA was also involved in the implementation of an anti-corruption project launched in 2005 with USAID funds (USAID 2006b). Located in the Amazonian city of Iquitos, the NGO has focused its activities on lobbying government for enhanced environmental regulations and implementing local sustainable development projects. While SPDA has often criticized the operations of transnational corporations, particularly in terms of intellectual property rights, it has not linked its concerns to a larger critique of Peru's political economy. Both SPDA and ACL are also limited by their direct strategy of engagement with the state.[19]

In terms of other Canadian organizations, Rights and Democracy provided support to human rights organizations such as *Asociación Pro Derechos Humanos* (Pro-Human Rights Association – APRODEH) in the 1990s and was a major supporter of Peru's Truth and Reconciliation Commission.[20] In more recent years, most of its support has been channelled to the *Centro de Culturas Indígenas del Perú* (Centre of Indigenous Cultures of Peru – CHIRAPAQ), an NGO dedicated to promoting indigenous culture and rights. Among other things, the organization helped launch a permanent forum for Andean and Amazonian women to promote the leadership of indigenous women. Its activities focus primarily around organizing workshops, though at times it also makes direct political demands upon

---

19   Sanborn et al. (2007), for instance, cite one case in which a company called Enersur established a thermoelectric plant in a zone of Ilo vigorously opposed by the organization. Enersur then failed to implement an environmental management plan agreed upon by the company, ACL, and the central and local governments until six years after the plant began operations.

20   As mentioned in Chapter 2, Rights and Democracy also published a report (Rousseau and Meloche 2002) calling upon the Canadian company, Manhattan Minerals, to recognize the legitimacy of the Tambogrande municipal referendum which confirmed the overwhelming popular opposition to the company's plan to develop a gold mine in the small town.

the state. In the 2006 elections, for instance, CHIRAPAQ (2006) and the network of indigenous women called upon García and Humala to integrate eight demands into their platforms. These demands focused primarily upon cultural recognition issues and the inclusion of indigenous women in decision-making processes.

Development and Peace's Peru Country Program contributed approximately $1.71 million to Peruvian civil society organizations in its 2003–2006 phase (Development and Peace 2006a), and an additional $2.67 million for its five-year programme from 2006 to 2011 (Development and Peace 2006b). In both phases of the project, the same 14 grassroots organizations have received support. As in Bolivia, the main objective of Development and Peace's programme in Peru has been to strengthen the capacity of social actors to promote change. The organizations supported by Development and Peace include several aligned with the most progressive factions of the Catholic Church which have traditionally championed human rights and social justice issues. While Development and Peace also supports some fairly mainstream NGOs, it provides support to grassroots organizations such as the federation organizing popular soup kitchens, the *Federación de las Mujeres Organizadas en Centrales de Cocinas Populares* (Federation of Women Organized in Popular Kitchens – FMOCCP), as well as civil society networks that promote an alternative economic vision, such as the *Grupo Red de Economía Solidaria del Perú* (Network of Economic Solidarity of Peru – GRESP). The latter organization brings together several NGOs and religious groups with the primary purpose of advocating the social economy as an alternative to the neoliberal model. GRESP is a signatory of the Havana Declaration (Red Intercontinental de Promoción de la Economía Social Solidaria 2007), for instance, which denounces neoliberalism and supports the ALBA as an alternative approach to trade.

Development and Peace's support to the *Comisión Episcopal de Acción Social* (Episcopal Commission for Social Action – CEAS) exemplifies its more grassroots approach compare to other North American actors that tend to support issue-based NGOs with weak social constituencies. As the social institution of the Catholic Church, CEAS is associated with liberation theology, which has its roots in the philosophy of the Peruvian priest, Gustavo Gutiérrez. The commission is composed of 'social pastors' from each of the country's diocese to support local bishops on social and economic issues. Unlike many of the NGOs that were associated with the human rights movement, CEAS has always been closely linked to trade unions, peasant groups and social movements, and played a leadership role in organizing a national chapter of the anti-debt movement, Jubilee 2000, which has called for a complete rupture with the neoliberal model.[21]

---

21   As an institution of the Church, however, the CEAS also confronts the resistance of more conservative forces, particularly those bishops aligned with the Catholic organization, Opus Dei, which has a strong following in Peru. Development and Peace has been criticized by conservative forces in the Catholic Church in Peru for funding pro-choice groups (Swan 2009).

Yet democracy assistance from Canada in general has targeted issue-based NGOs which have not engaged in a process of mobilization to challenge the prevailing social order. In this respect, even those NGOs which have advanced a more radical agenda have not deviated too far from the boundaries imposed by donor priorities and mainstream developmental thinking. Moreover, the Harper government's recent announcement that it would be providing several million dollars in funding to the Christian NGO, World Vision, to partner with Barrick Gold to 'increase the standard of living of 1,000 families affected by mining operations' (CIDA 2011) suggests that the Canadian state is now fully subordinating its development assistance to the requirements of Canadian capital as part of the significant changes currently restructuring the democracy assistance field of practice (discussed in Chapter 2). Additionally, according to the representative of a Canadian NGO who met with CIDA officials in Lima, the local CIDA office will be managing a \$5 million discretionary fund for organizations implementing projects related to corporate social responsibility.[22]

Coupled with the larger trends discussed in Chapter 2, it seems unlikely that Canada's grassroots tradition to democracy assistance will withstand the new offensive as more compliant NGO networks vie for state funds. Indeed, Development and Peace has been an outspoken critic of Canadian mining practices, and the new climate will most likely affects its activities in Peru. In 2006, it submitted a petition with over 150,000 signatures to Deepak Obhrai, parliamentary secretary to Maxime Bernier, Canada's Foreign Affairs Minister, urging the federal government to deny support to Canadian mining companies who violate international environmental and human rights standards (Mining Watch Canada 2006). World Vision, for its part, which speaks about empowering the poor and the oppressed, will provide new hegemonic resources for both the Peruvian state and Canadian capital while reinforcing Canadian foreign policy.

In a letter to World Vision-Canada, Barrick Gold and the Canadian Minister of International Cooperation, Miguel Palacin Quispe, the leader of the *Coordinadora Andina de Organizaciones Indígenas* (Andean Coordinator of Indigenous Organizations), criticized the track record of Canadian mining companies, stating that: 'no 'social works' carried out with the mining companies can compensate for the damage done, particularly in the face of rights having been violated'. Instead, Quispe called upon Canada to take 'responsibility to ensure that Canadian companies respect, and demand that States respect, the rights of the indigenous peoples affected before anyone seeks mining concessions in our countries' (Arnold 2012).

---

22 Interview, Ottawa: 22 March 2012. The interviewee spoke on condition of complete anonymity.

US Democracy Assistance under García

A New Hegemonic Discourse

The Canadian approach to democracy assistance remained fairly consistent throughout the period under question. Not so for the United States. As Toledo's term came to an end, the effort to create a new form of inclusive neoliberalism as a bulwark against both internal discontent and the left turn occurring throughout the region was increasingly unsuccessful, and US programmes began to advance a new repertoire of tactics as the balance of class forces began to shift. Although urban-based movements remained weak, indigenous-peasant movements were growing considerably in strength. In the Andes, the *Confederación Nacional de Comunidades del Perú Afectadas por la Minería* (National Confederation of Communities Affected by Mining – CONACAMI) mobilized Aymara and Quechua peasants against the destructive practices of the mining and extractive industry (Poole 2010). In the Amazon, the *Asociación Interétnica de Desarrollo de la Selva Peruana* (Inter-ethnic Association for the Development of the Peruvian Jungle – AIDESEP) became increasingly organized and confrontational with both mining companies and the Peruvian state (Hughes 2010). Despite the US-sponsored public education programme against them, coca growers were also becoming more militant and organized, though they remained internally fractious and mostly insulated from other social movements (Bebbington, Scurrah, and Bielich 2010). Peru's main union central, the *Confederación de Trabajadores del Perú* (General Confederation of Workers of Peru – CGTP) also engaged in militant actions, particularly the teachers' union, the socialist *Colegio de Profesores*.

In November 2005, Humala launched the PNP, a party whose base consisted primarily of the rural veterans of the conflicts of the 1980s and 1990s. Inspired by Peru's popular history of anti-imperialism and the socialism of its most famous Marxist intellectual, José Mariátegui, the PNP articulated a neo-developmentalist economic programme that called for an expansion of the internal market and the reregulation of the neoliberal economy. Aligning himself with other leftist leaders in the region, particularly Evo Morales and Hugo Chávez, Humala promised to transform the state through a radical-ethical vision of democracy.

Leaked cables from the US embassy in Lima indicate that US representatives were monitoring the situation very closely, meeting with local officials in areas such as Puno where PNP support ran high to assess the party's support base. The US embassy was also worried about the appeal of the *Colegio de Profesores*, as well as indigenous groups. Embassy representatives attended events such as the national *cocalero* congress in 2005, reporting back confidently on the many schisms between the movement that would mitigate its ability to influence the political situation (and the US-backed eradication plan) (Wikileaks 2005a). As for Ollanta Humala, one cable denounced him hysterically as 'fascistic' and a 'brown shirt', noting that his narrow appeal would likely dampen the ability of Chávez to export his Bolivarian Revolution to Peru (Wikileaks 2005b).

As the 2006 presidential election approached, Secretary of Defense Donald Rumsfeld visited President Toledo and the president of Paraguay to urge them to work together to stem 'antisocial, destabilizing behaviour' in a not so subtle suggestion that Chávez was providing financial support to Humala's PNP (Rumsfeld was greeted in Lima by throngs of protestors chanting 'Murderer! Murderer!' and shouting expletives as a military band played the Star-Spangled Banner (Associated Press 2005)). Toledo also spoke out against Chávez, accusing him of intervening in the 2006 election, and calling on him to 'learn to govern democratically'. Chávez, for his part, announced his hope that Humala would win the election in response to criticisms from García (Forero 2006).

The near victory of Humala in the 2006 presidential elections provided USAID and US democracy promotion in general with a new sense of urgency.[23] A USAID official interviewed in Lima openly affirmed that the main objective of the agency was to ensure that Humala did not win the next election. Displaying a map of Peru which depicted the positive correlation between the poorest regions and electoral support to Ollanta Humala in the 2006 presidential elections, the official warned that a 'populist candidate like Humala can repeat a message that resonates with the 40% of the population that has not really seen their lives substantially improve over the last several decades, threatening to take the country in a dramatically different direction'. Although many democracy assistance programmes supporting the government and state institutions remained in place, a new focus in civil society and on political parties emerged.

To orient this new phase of programming, the USAID-funded Latin American Public Opinion Project (LAPOP) released its first report on Peru in 2007 since 1998. As part of the Americas Barometer initiative, the cross-national LAPOP measures citizen perceptions on a plethora of issues, enabling USAID to draw upon this information to more strategically focus on specific issue-areas within its overall agenda of supporting liberal democracy and free markets. The report on Peru linked the rise of 'populism' in the 2006 presidential elections to crime, corruption and problems of political representation (Seligson and Carrión

---

23   There is some indication that NGOs which received US funding such as members of the CNDDHH may have contributed to a smear campaign against Humala by accusing him of being responsible for war crimes alleged to have taken place when he commanded a counter-insurgency base in the jungle in 1992 (Bigwood 2006) (these allegations resurfaced in 2011 but have never been proven). In the second round of the 2011 presidential elections, however, the CNDDHH actually mobilized against the candidature of Keiko Fujimori, a position tantamount to endorsing Humala. Most NGOs, however, did not explicitly align themselves with elite political leaders or parties. In fact, in the years that followed the election, several human rights NGOs – including CIDA and USAID-funded APRODEH – targeted the García government for its criminalization of social protest. In response to widespread criticism, the García government passed a law establishing a state institution to regulate NGOs – the *Agencia Peruana de Cooperación International* (Peruvian International Aid Agency) – and announced an investigation into the activities of human rights organizations (Alonso Ramos 2008).

2007). At a CIPE conference in Lima in September 2007, the USAID Assistant Administrator for Latin America and the Caribbean, Paul Bonicelli (2007), cited the dismaying results of the Americas Barometer surveys, emphasizing the importance of countering populist appeals by dealing with the very issues emphasized in the LAPOP report.

Accordingly, USAID began providing more support to NGOs dealing with issues of transparency as part of a discursive re-framing of the barriers to democratization away from human rights issues to those of corruption. As Sum (2008) argues, the production of hegemony is an ongoing process that often requires the remaking or fixing of 'common sense' around particular issues. As new issues emerged organically surrounding Peruvian democracy, the US recursively selected and magnified hegemonic discourses around transparency in an effort to reduce populist appeals and secure a discourse that was structurally coherent with its economic interests. The objective was clear; steer the public discourse on democracy in a non-radical direction.

One of the main organizations to receive US support was the *Consejo Nacional para la Ética Pública* (National Council for Public Ethics – PROÉTICA), which was established as a coalition of four civil society organizations dedicated to anti-corruption issues in 2001. In 2003, the coalition became the Peruvian chapter of Transparency International. PROÉTICA has engaged in numerous anti-corruption campaigns over the years and publishes a monthly report on corruption issues. USAID support to the coalition began through the OTI programme, but was considerably enhanced through an $850,000 grant in 2008 to build the capacity of civil society organizations to fight against corruption in the context of decentralization. Additionally, the coalition was involved in the IRI's political party strengthening programme (2008–2009), served as the implementing organization of a programme sponsored by the NED to fight regional corruption (2007–2008), and received funding from George Soros' Open Society Institute to strengthen regional anti-corruption institutions and civil society monitoring (2006–2007). PROÉTICA's founding organizations, particularly *Transparencia* (Transparency), have also been major recipients of US democracy assistance funds.

Based in Lima and led by urban professionals who opposed the Fujimori regime, PROÉTICA and its affiliated NGOs are watchdog groups that define their activities primarily in terms of monitoring public authorities. While this function is important, these organizations contribute to a limited vision of democracy that ultimately coincides with US interests. One of the organization's key partners and a founding member of the coalition, *Transparencia,* for instance, espouses a vision of democracy that is restricted almost exclusively to the political realm where it is concerned primarily with matters of efficiency. *Transparencia* and PROÉTICA show little concern with substantive issues of democratic citizenship and are generally uncritical of the deleterious effects of the state's economic project.

Moreover, both the *Comisión Andina de Juristas* (Andean Commission of Jurists – CAJ) and the Venezuelan chapter of the *Instituto Prensa y Sociedad* (Institute Press and Society – IPYS) – two other founding members of PROÉTICA

– have been vocal critics of Chávez, a position for which they were denounced by the Venezuelan government as puppets of the Bush administration. The US has explicitly funded such organizations as a strategic response to growing populism in the region (Wikileaks 2005b). Yet, none of them have played an active role in denouncing Humala and the PNP. *Transparencia*, which observed the 2006 elections, congratulated both Humala and García for their conduct throughout the electoral process. Likewise, the organization did not mobilize against Humala in 2011. While these NGOs have certainly contributed to a particular hegemonic framing of democracy which has been disseminated throughout civil society, they have not aligned themselves with right-wing elite social forces.

Strengthening Parties and Incorporating the PNP

In addition to the new focus in civil society, the United States began more actively targeting political parties in an effort to both undercut and co-opt the PNP in the wake of the 2006 elections. According to a second USAID official interviewed in Lima, the agency was particularly concerned by the rise of new anti-systemic movements at the regional level aligned with the PNP.[24] Accordingly, USAID funded the IRI's Political Party Strengthening Program (2008–2009) to counter-balance the rise of anti-systemic parties at the regional level by building stronger links between regional movements and national parties. The initiative included PROÉTICA, *Transparencia* and two other NGOs as project partners.

NDI also began working on political party strengthening after having completed a study in 2004–2005 on the political party system and pro-poor reform. The report, which led to a series of follow up programmes funded by the NED, highlighted the need to build the policy capacity of elected officials and political parties to articulate and implement post-Washington Consensus-style pro-poor policy reforms (NDI and the Department for International Development 2005), thereby enhancing their capacity to advance appealing policies without addressing underlying issues of class and inequality. In a report before the Congressional Subcommittee on Western Hemisphere Affairs in March 2005, Kenneth Wollack (2005), the President of NDI, described the de-politicization function of political party strengthening programmes in Peru in no uncertain terms. Wollack noted that: 'Civil society activism without effective political institutions quickly creates a vacuum. It sows opportunities for populists and demagogues who seek to emasculate parties and legislatures, which must serve as the intermediaries between the state and citizens and, therefore, are the cornerstones of representative democracy. This dangerous trend has already been seen in several countries in the Andean region – including Bolivia, Ecuador and Venezuela.'

Yet political party strengthening programmes have also included the PNP as participants. One member of the PNP's political committee who coordinated

---

24   Interview, Lima, 8 January 2009.

the party's involvement in such programmes noted in an interview that the vision of democracy being promoted is not favourable to a process of social change that would benefit the masses, but that US programmes had largely been beneficial to the party nonetheless.[25] Given the explicit aim of USAID and NDI to counter anti-systemic and populist tendencies in Peru, the fact that political party strengthening programmes have included the PNP represents a paradox. One plausible explanation for the incongruity is the nature of the PNP itself, which lacks a grassroots social base. Inviting the PNP to participate in political party strengthening programmes might serve the objective of bringing them into the political system through workshops that encourage a technocratic approach to dealing with policy issues. In this sense, a strategy of incorporation could help transmit hegemonic discourses that do not challenge underlying class relations.

Another factor is the need for caution in light of the global backlash against US democracy promotion, as discussed in Chapter 2. The crisis of legitimacy has been particularly acute in South America, where leaders such as Evo Morales and Hugo Chávez have publicly denounced US programmes. In this context, USAID may have opted for a more cautious approach of bringing the PNP into the system rather than organizing the traditional political parties against it.[26]

Humala Wins, but ...

The attempt to stem the tide of discontent during García's presidency failed, and Humala managed to win the 2011 presidential election, which occurred in the context of a new indigenous militancy. Regional strikes over mining, hydrocarbon concessions, and hydroelectric projects in Arequipa, Ancash, Cajamarca, and Moquegua coincided with the run-up to the first-round election on April 10 (Poole and Renique 2011). In Puno, the Aymará-led Natural Resources Defense Front mobilized in early May with support from the CONACAMI to demand the revocation of the Santa Ana silver mining concession granted to Canada's Bear Creek Mining Company. Although the protests were temporarily suspended for the runoff election on 5 June, they resumed soon after it and spread to Puno's northern provinces, where a Quechua-led movement mobilized against mining operations and the construction of a massive hydroelectric dam (Achtenberg 2011).

The election itself witnessed the split of the right among four candidates – Keiko Fujimori, Luis Castañeda (former mayor of Lima), Pablo Kuczynski (a former finance minister under Toledo), and Toledo himself. Leftist political parties

---

25   Interview, Lima, 17 January 2009.

26   While Canada does not officially engage in this area of democracy assistance, it contributed $97,239 to International IDEA to support greater political inclusion of women from 2005–2008 (CIDA 2009). Although IDEA often collaborates with NDI and IRI, I have not uncovered evidence to suggest that the multilateral organization is advancing a specific political agenda.

formed a broad coalition under the leadership of Humala, *Gana Perú* (Win Peru), which received the support of the indigenous movements (the AIDESEP and the CONACAMI) as well as organized labour (the CGTP). But as part of his campaign strategy Humala significantly moderated his discourse to appeal to a broader segment of the population, promising to leave the market untouched while making small changes such as increased taxes on foreign enterprises to finance more generous social policies. Humala also issued a 'Promise to the Peruvian People' that he would not seek re-election and would respect property rights and disavow nationalization. He publicly shunned support from long-time regional ally Hugo Chávez, bringing in political advisers from Brazil and presenting himself as a socially aware but fiscally conservative disciple of Lula (Farnsworth 2011). Many middle- and upper-class voters reluctantly rallied behind Humala as a lesser evil than Fujimori's *Fuerza 2011* (Force 2011) in the second round of voting.

In his first month in office, Humala removed any lingering suspicion that he would break with the policies of the previous two governments by naming Luis Miguel Castilla, a technocrat and former deputy minister of finance under García, as minister of the economy and finance. He also reappointed the orthodox economist Julio Velarde as Central Bank president. As Achtenberg (2011) notes, Humala's cabinet combines market-friendly economic managers with left intellectuals in charge of social programmes. Although he has made good on some of his social promises by raising the minimum wage and pensions, he has yet to increase taxes on mining companies to pay for the increases as originally planned. At best, the new social policy may attenuate poverty and perhaps even modestly reduce Peru's glaring inequalities. The government's heavy-handed response to a new round of protests in the region of Cajamarca against a proposed $4.8 billion mine by US-based Newmont Mining suggest that even a modest post-neoliberal agenda may be unlikely. Humala quickly dismissed the protestors' environmental concerns against the mine – which would be the largest in the country's history – declaring a state of emergency and replacing his prime minister with a former military officer known for his confrontational approach (Wade 2011).

In the end, the United States had very little to fear after all – ironically, Humala may be the best thing yet in the effort to stabilize neoliberal polyarchy in Peru. Humala's accumulation strategy will continue to privilege Lima's business elite and agro-mining oligarchy, which have historically shown little interest in making Peru more socially just or democratic. If progressive social forces are unable to pull Humala farther left, he will most certainly adopt a pragmatic approach in keeping with the inclusive neoliberalism pioneered by Toledo. Their ability to do so will likely depend upon whether they can consolidate a truly national popular movement. With orthodox neoliberals in key positions, Humala himself is unlikely to play a leadership role as Chávez has in Venezuela in further developing the collective power of the left. In the chimerical effort to secure a new equilibrium between Peru's dominant and subordinate classes and ethnic groups, he is more likely to govern by attempting to co-opt the social movements as have the Kirchners in Argentina and, arguably, Evo Morales in Bolivia (both of whom have only been partially successful). The ongoing cycle of protests indicates that this will not be an easy task. With the neoliberal state firmly ensconced, as North American programs had intended, the impetus and momentum for change will continue to come from below.

Concluding Remarks

The case of Peru demonstrates the need to expand the critique of democracy promotion to theorize how it contributes to elite hegemony through more subtle forms of social control in countries where North American allies are in power. In contrast to Haiti, where there has been intense political polarization, elite political coalitions in Peru have dominated political society with little organized opposition. In such circumstances, democracy promotion and democracy assistance have contributed to the state's project of passive revolution. But US and Canadian approaches also varied more so than they did in Haiti. For the United States, the focus during the Toledo years was in building the legitimacy of the government, supporting new forms of limited inclusion, and funding civil society organizations promoting a liberal-democratic political discourse. Canada also sought to stabilize the political situation, but its approach was less focused on Toledo's government and more focused on strengthening the state as a whole. As a result, Canadian programmes remained fairly consistent throughout the Toledo and García presidencies. Canadian organizations such as Rights and Democracy and Development and Peace have also supported civil society actors with a wider social base that have called for a genuine process of social change.

With the rising tide of regional discontent and the near victory of Humala in the 2006 presidential elections, US programmes began to focus on rebuilding political parties and re-framing the democratic discourse around issues of transparency. But it also chose to include the PNP directly in its capacity-building efforts. Unlike Lavalas, the PNP lacked a grassroots base and the United States may have considered a strategy of co-optation the more prudent course of action. The shifting regional balance of power also likely had a strong affect since the crisis of legitimacy of US democracy promotion was particularly acute in South America, where leaders such as Morales and Chávez publicly denounced US programmes. As we shall see in the following chapter, such public denunciations seem to have put an end to a strategy of regime change against the MAS around the same time. USAID may have therefore opted for a more cautious approach of bringing the PNP into the system rather than organizing the traditional political parties against it.

Those North American programmes that sought to stabilize the situation ultimately failed. Despite minimal social concessions and institutional mechanisms for inclusion, political and economic elites remain generally averse to any form of intervention by popular social groups in state life. The Peruvian state has not produced an integral hegemonic order based on some degree of class compromise, and a decadent or minimalist hegemony has prevailed. In this context, the main challenge ahead for Peru's revitalized social movements will be to forge a truly national popular movement with an alternative to neoliberalism.

Chapter 5
End Game in Bolivia

Over the last decade, left indigenous social forces in Bolivia have led a series of revolts that have contested the very legitimacy of the state. The failed attempt by successive governments to create a new hegemonic order combining neoliberalism with new mechanisms of inclusions provided the backdrop against which a militant indigenous popular movement led by the *Movimiento al Socialismo* (Movement towards Socialism – MAS) came to power in 2006, with Bolivia quickly joining the ranks of the more radical governments in the region led by Venezuela. Since then, however, Bolivia has been divided internally between two competing hegemonic blocs – an indigenous popular bloc in the west led by the MAS against a predominately white oligarchy with middle class support led by departmental prefects in the wealthy eastern lowlands.

Among other things, the MAS have confronted the traditional imperial politics of the United States, including its democracy promotion programmes. Bolivia's vice president, Álvaro García Linera, summarized the conventional wisdom succinctly: 'Wherever there is conflict, if you dig a little, USAID or NGOs linked to USAID are at the heart of the problem' (Paredes 2010). US interference in Bolivian politics is nothing new, and there is considerable evidence that its democracy promotion programmes were configured to support social forces hostile to the MAS as well as NGOs and civil-society organizations that posed as moderate alternatives to it (though the extent to which the party has actually broken from neoliberalism is a matter for debate).[1] While the Bolivian government has consistently denounced US democracy assistance programmes as one facet of its overall interventionist foreign policy, Canada has yet to suffer the same critique. This chapter examines the tactics associated with US and Canadian democracy promotion programmes in Bolivia from before the second presidency of Gonzalo ('Goni') Sánchez de Lozada in the early 2000s until the completion of Evo Morales's first term in office in 2009. During this time, US democracy promotion adopted a mixture of hard and soft tactics in response to the shifting balance of political power. Programmes seeking to undermine the MAS have been well documented in the critical literature (see Allard and Golinger 2009, Lindsay 2005, Beeton 2009, Dangl 2008, and Bigwood 2008), but few works have looked at the way hard and soft tactics have interacted and responded to the evolving political conjuncture, and none have included Canada in the analysis.

---

1   See Webber (2011a, 2011b) and Petras and Veltmeyer (2009) for a radical critique of the MAS, and Fuentes (2011) for a socialist defence.

The analysis is divided into three parts. The first part sets the historical context through an analysis of the state and competing social forces over the last decade, as well as US and Canadian relations with Bolivia. I then examine US and Canadian programmes prior to the Morales victory in 2006. During this period, both Canada and the United States sought to stabilize the neoliberal state through tactics similar to those analysed in Peru. This included reinforcing the legitimacy of state institutions in an attempt to respond to the mounting crisis through passive revolution. As in Peru, moderate NGOs articulating hegemonic discourses on democratic compromise and restraint were crucial to these efforts. Canada again focused primarily on building the legitimacy of the democratic system as a whole whereas the United States was more concerned with directly supporting the government.

I then examine the US shift to a strategy of regime change through a decentralization programme designed to empower regional prefects as a counterweight to the MAS and left indigenous social forces in the eastern lowlands. Soft tactics – including programmes supporting moderate civil society organizations and party-strengthening programmes that actually included the MAS as participants – complemented these efforts. I also argue that the United States was constrained in its ability to transition to a full strategy of regime change as a result of the growing backlash against US democracy promotion in the region. I then juxtapose US efforts in civil society with the local alliances that were forged by Canadian NGOs, which in many cases supported the struggles of popular civil society. This focus provides further empirical evidence that the grassroots tradition was still present in the 2000s despite the extent to which it was compromised in Haiti (though, as argued in Chapters 2 and 4, it is currently being replaced by a network of more compliant NGOs working in collaboration with Canadian mining companies).

The conclusion explores the MAS's response to US democracy promotion, which has been characterized by a combination of resistance and accommodation. While it has refrained from shutting down US programmes, the Morales government has been an important voice in the regional backlash against US democracy promotion, and in constraining its ability to wage campaigns of destabilization against left governments in the name of democracy. Bolivia may not signify the end of the democracy wars, but its experience does point to the more limited terrain and range of tactics with which they may be fought.

Competing Hegemonic Projects and the North American Presence

Bolivia's geographic diversity is nearly as impressive as Peru's, though it has remained a landlocked country for most of its history. As the Andes descend to the east from the highland region around La Paz, the fertile agricultural regions of the sierra lowland gradually give way to Bolivia's share of the Amazonian rainforest. One of the most indigenous countries in Latin America, 55% of Bolivia's population of 9.52 million inhabitants is of indigenous descent, 15% European, and 30% mestizo (Central Intelligence Agency 2012). The vast majority of people of European and mestizo descent are concentrated in La Paz – the seat of the executive and the legislative branches of government – and the prosperous agricultural regions of the eastern lowlands (Sucre is the constitutional capital, where the Supreme Court is located). The indigenous population is concentrated in the high plateau around Lake Titicaca, or altiplano, of the western highlands (including around La Paz), and is divided mainly between the Aymara and Quechua ethnic groups. Bolivia is extremely rich in hydrocarbons, natural gas, and mineral deposits, with significant petroleum, zinc, iron, lead, gold, timber, silver, and tin reserves. The extreme inequality between class and ethnic groups has led to a history of militancy amongst its miners and peasants. In 2007, the richest 10% of Bolivians held 45% of all income while the poorest 20% held a mere 2.6% (World Bank 2012).

Both the United States and Canada have growing economic ties with Bolivia, particularly in the mining and hydrocarbons sector. Bilateral trade relations with the United States have increased over the years, though they are still relatively small: in 2000, the United States exported $253 million to Bolivia and imported $184.9 million in return; by 2008, these figures had increased to $389.3 million and $511 million respectively (US Census Bureau 2010c). The United States is the largest source of FDI in Bolivia, however, accounting for approximately one-third of net FDI inflows of $1.7 billion between 2001 and 2006. From 2002 to 2008, US companies invested an estimated $750 million in the mining sector, USD $420 million in the hydrocarbons sector, $290 million in energy production and distribution, and $230 million in telecommunications (US Department of State 2008b). Canada's trade relations with Bolivia are modest: in 2007, for instance, exports to Bolivia totalled $15.2 million while imports equalled $106.4 million. At the same time, Canadian direct investment in Bolivia in 2006 amounted to $87 million (Government of Canada 2008) and Canadian companies held the dominant share of the larger-company mineral exploration market in 2007 (Natural Resources Canada 2009). As illustrated in Figure 5.1 below, both Canada and the United States spent considerable amounts in democracy assistance in the first half of the 2000s. From 2001 to 2006, Canada spent $30 million while the United States spent $51 million. Both countries increased their assistance as the neoliberal state increasingly came under attack.

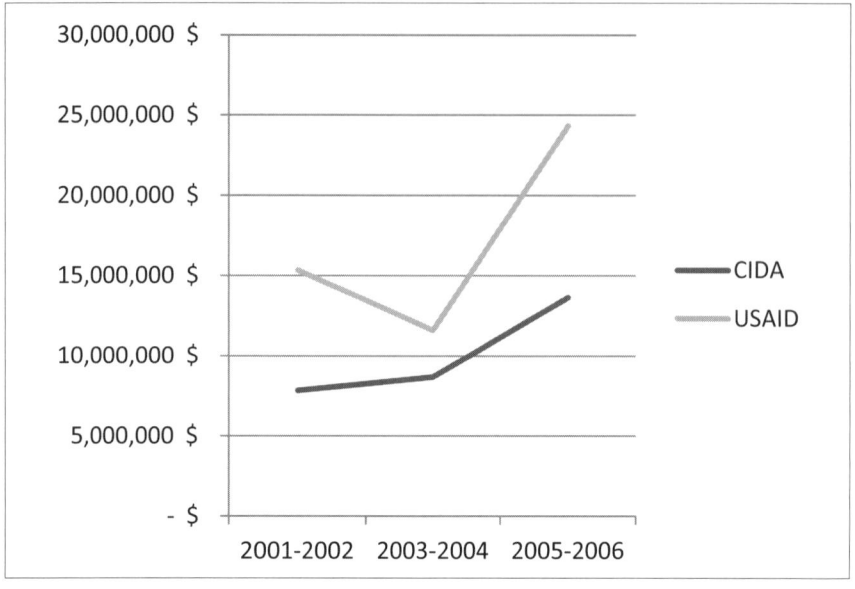

**Figure 5.1 US and Canadian democracy assistance in the lead up to the
Morales victory**

The victory of the MAS in the 2005 general and presidential elections marked a
significant rupture with the neoliberal state that emerged in the wake of Bolivia's
transition to polyarchy in the early 1980s. After several cycles of neoliberal reform
and the implementation of draconian austerity measures in the 1980s, Goni's
Movimiento Nacionalista Revolucionario (National Revolutionary Movement –
MNR) sought to contain left indigenous social forces through reforms such as the
Law of Popular Participation (1994), a municipal decentralization that provided
for a certain degree of local self-governance of indigenous communities.[2] As
popular forces grew in collective power, however, Bolivia passed through a
'revolutionary epoch' characterized by large-scale regional revolts and a crisis of
authority of the neoliberal state (Webber 2011a).

   The Cochabamba Water Wars in the early months of 2000 marked the
beginning of a new cycle of revolt that shook the foundations of Bolivia's pacted
democracy. The new mobilization occurred in the wake of the electoral victories
of the MAS and other indigenous parties at the local level in the second half of
the 1990s. Thereafter, indigenous groups, landless peasants, workers, *cocaleros*,
women, teachers, and other popular sectors increasingly coordinated their

---

   2   A wealthy businessman who benefitted immensely from the privatizations of the
1980s, Goni spent so much time in the United States that he spoke Spanish with an English
accent.

oppositional actions, in effect forming a counter-hegemonic national popular bloc. The popular revolt culminated in the Gas Wars of October 2003, when citizens from El Alto blocked the main highway to La Paz after the government announced plans to export gas to Mexico and the United States through a Chilean port. After violent confrontations between protestors and security forces which left at least 59 people dead (Assies 2004), Sánchez de Lozada gave his resignation to Congress and was succeeded by his vice president, Carlos Mesa Gisbert. Mesa sought to deal with the protests through more consensual means, acquiescing to popular demands to establish a constituent assembly to rewrite the constitution. Congress stalled the constitutional project, however, and Mesa eventually resigned over his inability to control the state's increasingly repressive turn. Mesa had also faced the resistance of the IFIs and the United States, whose ambassador, David Greenlee, informed him that any reversion of the neoliberal economic plan and coca eradication could threaten US support (Kohl and Farthing 2006: 179). In the end, the confrontation between civil society and the state only ended with the 2005 presidential and general elections, which brought Evo Morales to power and gave the MAS a congressional majority.

The popular movement led by the MAS has called for a complete rupture with neoliberalism and the legacy of internal colonialism. It has encountered the resistance of a right-wing regional opposition exemplified by the prefects of the *media luna* (so named because the departmental capitals form a 'half moon' on the map) and the *comités cívicos* (civic committees), which represent the interests of white oligarchs and middle class mestizos in the eastern lowlands. The urban middle class of La Paz has also been critical of the MAS.

The main mechanism through which the government has sought to consolidate its social vision is the new constitution that was approved by popular referendum in January 2009. Since the theme of constitutional change is perhaps the most important topic to be addressed by democracy assistance agencies and local civil society organizations over the past five years, a few comments on how the process played out are in order. A constituent assembly first convened in August 2006 in Sucre based on the direct election of representatives associated with the political parties. From the outset, the MAS was criticized by many popular movements for entering into negotiations with the opposition in the Senate – where it did not hold a majority. This precluded the participation of civil society and provided a new institutional space for the remobilization of the opposition (Webber 2008). Despite this concession, the assembly quickly polarized along party lines. The MAS was heavily criticized for wanting to use a simple majority rather than a two-thirds majority formula to pass the constitution, and when protests against the MAS's constitutional project erupted on the streets of Sucre, the government moved the assembly to Oruro, which led to a boycott on the part of the representatives of the opposition parties. In the end, the new constitution was passed by the assembly in December 2007 with only 165 of its 255 members present.

The prefects of Beni, Pando, Santa Cruz and Tarija responded by organizing referenda on regional autonomy, all of which passed in May and June 2008.[3] When the regional governments moved to implement the autonomy statutes, the central government organized a recall referendum on the mandates of eight of the nine regional prefects as well the president and vice-president. Morales and Linera won the recall with 67% while two of the prefects, both aligned with the opposition, failed to garner enough support to prevent new elections. The conflict between the central government and the departments came to a head in August and September 2008 as MAS supporters clashed with regional authorities and members of the *comités cívicos* throughout the *media luna*. In Pando, 20 peasants were killed in a massacre organized by death squads in support of autonomy (Hylton 2008). Morales then declared a state of emergency in the department, ordering the arrest of the regional prefect, Leopoldo Fernández. After tense negotiations, Morales finally reached agreement with the prefects to hold a referendum on the constitution in January 2009, at which point the new *magna carta* was accepted by a 61.7% majority.

Democracy Promotion in the Lead Up to the Morales Victory

The United States Reinforces the Hegemony of the Inclusive Neoliberal State

As a strong ally of the previous neoliberal administrations, the United States has vociferously opposed the MAS government. This of course is not surprising given the MAS's political base in the *cocalero* and indigenous social movements, which have consistently opposed US coca eradication policies and the rights of transnational corporations. The use of both soft and hard tactics of democracy promotion to undermine the MAS are rooted in the historical pattern of 'eternal interference' in Bolivian affairs (Calloni 2009), which has long supported the shared interests of US foreign capital and the Bolivian financial and agro-industrial oligarchy. Since the 1960s, military and covert political aid to Bolivian conservative factions has helped to entrench patterns of economic exploitation and racial oppression of the indigenous majority.[4] The activities of US democracy promotion agencies are a continuation of these historical patterns of intervention by

---

3   Eaton (2007) argues that the strategy of advocating for regional autonomy was a response to the fact that indigenous actors had permeated the other two levels of governing authority – the municipal and national.

4   US support for reactionary political forces in Bolivia through the CIA helped lay the groundwork for the military coup of General René Barrientos in 1964, which reversed many of the gains achieved by the National Revolution in 1952 (Calloni 2009). These gains included the nationalization of Bolivia's mines and the breaking up of powerful estates. Under the dictatorship of Hugo Banzer Suárez (1971–1980), the United States consolidated its alliance with a new agro-industrial oligarchy based in the fertile eastern lowlands,

more acceptable means. They extend the low-intensity political warfare pioneered by the CIA and are closely intertwined with more coercive elements of US policy, including regular political interference by the US embassy and Drug Enforcement Administration (DEA)–sponsored coca eradication programmes.

USAID situated its Democracy and Governance programme in Bolivia from 1998 to 2003 within the strategic objective of obtaining 'increased citizen support for the Bolivian democratic system'. Thus, the approach was initially formulated under Clinton and taken over by the Bush administration. US support for the state included a programme intended to reinforce the legitimacy of the congress. Launched in the early 1990s and re-established in 2001 after a two-year hiatus, the programme aimed to improve communication between representatives and constituents through 'outreach mechanisms' such as public hearings, forums, and interactive radio programmes. Outreach events were sponsored by USAID in Bolivia's nine departments and were credited in one official report with building trust of governing institutions and reducing the risk that 'anti-system misinformation campaigns' would undermine political instability (USAID-Bolivia 2003: 3).

USAID also provided the transnational development firm Chemonics International with US$15 million from 1996 to 2003 to implement a programme that permeated the layers of municipal and state institutions and civil society. Bolivia's two national-municipal federations, the *Federación de Asociaciones Municipales* (Federation of Municipal Associations – FAM) and the *Asociación de Concejalas de Bolivia* (Association of Women Councillors of Bolivia – ACOBOL), all nine departmental associations of its municipalities, and some 50 grassroots organizations and 54 NGOs took part in the programme (Chemonics International 2003). Both FAM, which represents all of Bolivia's municipalities, and ACOBOL, an institutional mechanism representing women councillors and mayors within FAM, grew out of the decentralization initiative of the 1990s; USAID was also instrumental in setting them up. In the first phase of the programme, Chemonics developed a series of municipal governance methodologies with limited participatory features along with oversight mechanisms such as citizen vigilance committees. These mechanisms helped produce a 'watchdog' construct of citizenship reduced to ensuring the transparency of the state, particularly with regard to the financial management of budgets by municipal authorities (see Legoas 2007).

Chemonics was particularly concerned with creating the impression that democracy was working effectively through the decentralization plan, a concern that it, too, linked to the threat of anti-systemic forces. One report warned that the 2004 municipal elections would provide the MAS with the opportunity to bash ''the system' in general' and hoped that more effective mayors and councils would be put in place to re-establish confidence in Bolivia's democratic institutions (Chemonics

---

particularly in Santa Cruz de la Sierra, where land reform was barely carried out and traditional haciendas were mostly converted into extensive agro-businesses (Eaton 2007).

International 2003: 75). USAID also sought to co-opt MAS strongholds in the coca-producing regions of the Yungas and Chapare by bringing the municipalities into the national mainstream and building their capacity to provide services and resolve conflict (USAID-Bolivia, 2003: 8).

In civil society, IRI launched a programme in 2003 supporting the state's hegemonic agenda by countering dissatisfaction with political parties and 'increasing civic understanding of the role of the citizen' to 'channel civic interest into more constructive avenues of communication and participation' (IRI 2005c: 1–2). Funded by USAID, the Improving Citizen Perceptions of Political Parties programme coordinated 19 informational forums on topics ranging from the constituent assembly and the referendum to the role of indigenous groups in Bolivian politics. In addition to the forums, IRI conducted 27 radio programmes on the referendum and other issues and created a 10-part television programme on youth and democracy. Over 200 civil society organizations, including NGOs, teachers' organizations and labour unions, were involved in the project, as well as the Ministry of Education, the National Electoral Court, the departmental electoral courts, and even the Bolivian Association of Political Science. A high-school civic education initiative went so far as to develop a manual 'exploring the civic concepts of responsibility and authority' to complement a curriculum introduced by the Ministry of Education. Some 4,500 social studies teachers, nearly 90% of the instructors in this subject area, were trained in the use of the manual. The project sought to stabilize the political system and elite patterns of governance by disseminating discursive constructions calling for citizen restraint; not a single indigenous community, let alone authority, was mentioned in the manual (Lindsay 2005).

A USAID-funded civil society initiative carried out by the State University of New York's Center for International Development complemented IRI's work. In the first phase, the Support for Electoral Process in Bolivia programme oversaw a comprehensive campaign to promote the concept of representative democracy in the context of national and regional elections. Through the programme, local trainers and activists were mobilized to reach nearly 1,145,000 citizens through workshops and radio programmes that reinforced respect for the legality of representative democracy and discredited extra-parliamentary forms of political mobilization. A mass media campaign was particularly wide in scope, with 15 million television viewers and another 15 million radio listeners compelled through 'jingles and images' to think about the conditions necessary for representative democracy (USAID 2006c: 6). The attempt to create new liberal citizens in the context of what was happening can be interpreted as a strategy of de-radicalizing the MAS's social base while promoting liberal democracy as a hegemonic ideology. After the MAS had come to power, communications and civil society networks were used to warn Bolivians that the constituent assembly should not take on 'attributes of the current government', which would be risky for the 'peaceful resolution of outstanding regional and social conflicts'.

The focus on disseminating the state's hegemonic ideology through a range of capillaries across Bolivian civil society was propped up by an attempt to rebuild the traditional political party system. Since Bolivia's transition to polyarchy in 1982, the stability of the system had been predicated on shifting elite power-sharing arrangements among the three dominant parties – the MNR, the *Acción Democrática Nacionalista* (Democratic Nationalist Action – ADN), and the *Movimiento de la Izquierda Revolucionaria* (Revolutionary Movement of the Left – MIR). As the state's crisis of authority deepened, USAID funded a political-party reform project through NDI that, according to the US embassy, was intended to 'dovetail with the MNR's inclusiveness plank and, over the long-run, help build moderate, pro-democracy political parties that can serve as a counter-weight to the radical MAS or its successors' (Bigwood 2008). One NDI programme launched in 2002 brought 13 emerging political leaders from the traditional parties to Washington for seminars. New political parties such as the MAS and the radical *Movimiento Indígena Pachakuti* (Pachakuti Indigenous Movement – MIP) did not participate (Lindsay 2005). NDI also conducted research on pro-poor reform that informed a series of programmes funded by USAID and the NED from 2003 to 2005, focusing on building the capacity of political parties to increase transparency and reach 'underrepresented sectors' (NDI 2005). Such support helped reinforce party platforms through the inclusion of limited social demands reflecting the logic of inclusive neoliberalism. A second NED-funded programme, which ran from 2005 to 2006, sought to bridge political divisions in Bolivia's most contentious regions – La Paz, Santa Cruz, and Cochabamba – through far-reaching support across civil society, particularly in youth-led sectors.

An NDI representative in La Paz pointed out that support for multiparty initiatives and civil society organizations is intended to overcome ideological polarization and bridge social divisions.[5] But such support also builds the institutional strength and leadership of elite parties whose traditional response to the class conflict is a mixture of coercion and passive revolution. It works against the interests of the MAS, the most popular and disciplined political party, by building up political forces that can contest its growing hegemony. It is also intended to counter populism in the region, as the president of the NDI, Kenneth Wollack, made clear in 2005 before the congressional subcommittee responsible for Western Hemisphere Affairs (discussed in Chapter 2). IRI also launched a political-party-strengthening programme in response to 'the social unrest and subsequent democratic crisis' unleashed by the constituent assembly.

NDI and IRI also collaborated with International IDEA in the implementation of a $700,000 USAID programme, Support to Democratic Processes, which ran from October 2007 to June 2008 and focused on rendering political parties more transparent and internally democratic.[6] There is evidence that the gender-

---

5   La Paz: 20 February 2009.

6   See USAID Bolivia web site at: http://bolivia.usaid.gov/CurrentPrograms/DEM/06_NDI_IRI.pdf.

related activities of IRI and NDI, as well as the multilateral organization, the International Institute for Democratic and Electoral Assistance (International IDEA), reinforced a key political coalition against the MAS through their support to the *Unión de Mujeres Parlamentarias de Bolivia* (Union of Women Parliamentarians of Bolivia – UMPABOL), which brings together women parliamentarians. By IDEA's own admission,[7] UMPABOL is dominated by women from the opposition parties and has become a strong opponent of the government.[8] Although it is unclear as to whether all of these agencies have supported UMPABOL as a deliberate strategy of undermining the MAS, it constitutes yet another example of how democracy assistance can alter the balance of power in political society away from popular forces.

With the exception of the NDI's first programme, however, both NDI and IRI have largely taken a soft, multiparty approach to building elite hegemony through programmes that have – since 2004 – included representatives of the MAS. Indeed, defenders of democracy promotion in Bolivia regularly point to the fact that the MAS has been included in specific initiatives in attempt to down play their political objectives. One adviser with IDEA noted that not only do the MAS participate in its projects but it actually has the highest quota, since resources are assigned on the basis of the representation of the party not only in parliament but also in the municipalities. USAID-funded research has also emphasized the importance of drawing in MAS leaders prepared to work within the system and achieve their objectives through electoral means (for example, Gamarra 2003). The growing backlash against US democracy promotion likely also encouraged a more cautious approach than that which was seen in Haiti, where Lavalas had been excluded from IRI's programs (though the party did participate in NDI's).

Canada Concedes

Canada's approach to supporting democratic development in the realm of political society has been similar to its approach in Peru, focusing, in particular, upon supporting democratic institutions. There has also been considerable continuity between the Liberal governments of the early 2000s and the Conservative government of Stephen Harper. This section will summarize such initiatives, which are much more modest in scope than their US counterparts. However, the contribution of Canadian actors to supporting civil society organizations will be discussed in the final section by way of contrast with the approach of the United States.

Canada's democracy programmes were situated within its *Development Programming Framework*, which ran from 2003 to 2007. A document summarizing the framework praised Bolivia's neoliberal reforms while recognizing that

---

7   Interview with IDEA representative. La Paz: 19 February 2009.

8   In one statement, UMPABOL (2008) accused Morales with seeking to install a hidden dictatorship (*dictadura camuflada*) through violence and intimidation.

structural adjustment had had certain adverse effects. Expressing concerns over the fragmentation of the traditional party system, CIDA (2007) noted that the near victory of Morales in the 2002 presidential elections and the rise of anti-systemic parties such as the MAS in the Congress represent 'a very disconcerting trend to many Bolivians'. Consequently, the programme allocated $17 million of a total envelope of $50 million to democratic governance, focusing primarily on supporting state institutions.

This included $5 million to the *Defensoría del Pueblo* (human rights ombudsman) alone – making it by far the largest single recipient of Canadian democracy assistance. The *Defensoría* is one of Bolivia's most respected democratic institutions. Its mandate is twofold: defend the rights of Bolivians from abuses by state authorities and defend and promote human rights in the country more broadly. Canada was the first donor to support the institution (CIDA 2009). As in Peru, the Office of the Ombudsman has played an important role in advocating for the rights of marginalized and exploited sectors of the population. Although it may be viewed critically as a mechanism intended to manage and defuse social conflict while underlying issues are left unaddressed, this role does not preclude a more radical advocacy role. During the gas wars in the fall of 2003, the *Defensoría* strongly denounced the actions of the state when a confrontation between demonstrators in the town of Warisata and security forces left seven dead – an event that triggered the national mobilizations that led to Goni's resignation. This position contrasted with pro-government perspective of United States, which maintained that the actions of security forces in Warisata were justified. This position was stated by Ambassador David Greenlee during a ceremony in which the United States gave Bolivia $63 million in aid just as the government was on the verge of falling (Dangl 2007).

As the political crisis reached its apex and elections were scheduled for December 2005, Canada provided assistance to the national electoral commission, the *Corte Nacional Electoral* (CNE), to organize elections (CIDA 2007). Canada also committed approximately $50,000 to the OAS-led observation mission for the presidential, legislative, and provincial governor elections. The mission emphasized the overall integrity of the electoral process, concluding that 'we believe, without the shadow of a doubt, that the elections held in Bolivia on December 8, 2005 were peaceful, free, fair, and massively participatory' (OAS 2006: 44). The OAS report emphasized the fact that the elections witnessed the highest voter turnout (84.5%) since the transition to democracy in 1982 as well as the first absolute majority.

Canada's approach to stabilization during this period was thus to improve the overall functioning of democratic institutions rather than enhance the image of the government. While the US also supported both the *Defensoría* and the elections, its approach included a more explicitly political dimension. Again, this speaks to important differences in the ways Canada and the United States have gone about stabilization, and the fact that Canadian officials are less likely to act as political operatives. According to one high-ranking official at the Embassy section in Bolivia, Canadian interests in Bolivia have historically been minimal,[9] and democracy promotion has been carried out without significant interference from Ottawa. Following the presidential victory of Morales, moreover, Bolivia continued to be one of twenty priority countries for Canada's bilateral aid efforts. The official warned, however, that the Americas Strategy of the Harper government could lead to a more ideological approach, which he strongly cautions against given its failure in the case of the United States. For the time being, however, the Conservative government seems to be acting cautiously so as not to jeopardize its growing economic interests in Bolivia.[10] Such differences also likely reflect the US's more entrenched historical role in enforcing regional order. As we saw in Peru, Canada has yet to mobilize its status as a middlepower to pursue a more interventionist approach in the Andean region.

In the wake of Morales' victory, CIDA launched an $18 million project, Strategic Governance Mechanism, to run until 2012. The overall approach of the programme is designed to strengthen the technical capacity of state institutions, including the National Institute of Statistics, the Auditor General's Office, the National Electoral Commission, and the *Defensoría*, which continued to be the largest recipient of assistance. CIDA's Gender Equality Basket Fund channels additional assistance to a government – multi-donor basket fund which provides institutional support to the Vice-Ministry of Gender and Generational Affairs. From 2005 to 2008, CIDA contributed $1.1 million to the multi-donor basket fund (CIDA 2007). CIDA also contributed $500,000 to the OAS observation mission for the recall referendums in August 2008 (CIDA 2009l), which upheld, once again, the overall integrity of the process (OAS 2009).

Nonetheless, Canadian development programmes have been manipulated to stabilize the political situation in the interests of Canadian mining companies. Although CIDA does not classify its programmes in the hydrocarbon and mining sectors under the category of governance in Bolivia as it does in Peru, it is worth briefly commenting on them since they deal directly with strengthening state institutions. According to the report of the Auditor General of Canada, Canada has been influencing hydrocarbon policy in Bolivia since 1989, when CIDA, Petro-

---

9   For instance, until 2000 there was only one CIDA representative in La Paz – the Head of Aid, who was also charged with consular affairs. Canadian staff increased to two in 2002 and a third member was added in 2006 (CIDA 2008b).

10   Canada's aggressive rhetoric against Venezuela was contrasted with its more strategically tolerant approach towards Bolivia in Chapter 2. Nonetheless, some Canadian officials, such as former Ambassador to the United Nations and Liberal MP, Alan Rock, have been less discreet. In a lecture on reforming the United Nations, Ambassador Rock (2006) warned that some countries have become impatient with the economic mandates of the international financial institution (IFIs) and impatient for democratic dividends. Rock noted that: 'some of them have elected populist, socialist governments recently, as with Bolivia with Evo Morales, as in Venezuela with Hugo Chavez, and perhaps Peru, we'll see what the election results produce. And together with Cuba, those socialist, populist governments can cause quite a bit of mischief.'

Canada and the Bolivian government began working together to modernize the public oil and gas industry through the Bolivia Oil and Gas Project. The report notes that 22 Canadian firms received spin-off benefits from the project but does not question whether Canadian economic interests coincide with an inclusive development policy for all Bolivians.[11] In fact, the regulatory framework which CIDA contributed to has been even more of a focal point for popular mobilization than it has been in Peru. The gas wars that erupted in 2003 marked the culmination of popular discontent against this model. In the same year, CIDA launched a $13.25 million Hydrocarbon Regulation project, whose stated objective was to improve the regulatory framework to 'ensure sustainable resource development while maximizing benefits to Bolivia' (CIDA 2009). Although the programme has largely focused on providing technical assistance to the Ministry of Hydrocarbons and the state oil company, *Yacimientos Petrolíferos Fiscales Bolivianos*, the timing of the programme indicates CIDA's attempt to intervene on behalf of the state in opposition to the popular movements that were vigorously rejecting the way in which Bolivia's natural resources were being managed. While CIDA's lack of transparency[12] renders it difficult to evaluate the effects of the programme, it is clear that the implementing agency, IBM Business Consulting, favours a liberal approach to resource management.

A final area of Canadian democracy promotion pertaining to political institutions consists of the assistance channelled to Bolivia's two main municipal peak associations, also supported by USAID: FAM and ACOBOL. Through CIDA funds, Rights and Democracy and the Federation of Canadian Municipalities (FCM) – International have implemented capacity building initiatives with these organizations both separately and in collaboration. Beginning in 2002, FCM provided institutional-strengthening support to its counterpart, FAM, including expertise in policy development to enable it to act as a more effective interlocutor with the central government. Another important theme over the years has been strategic planning for local economic development. In 2008, Rights and Democracy and FCM launched a joint programme to support ACOBOL.

---

11   See Office of the Auditor General of Canada 1996 chapter 29, exhibit 29.11. This report was uncovered by Dawn Paley (2006) in a short exposé that appeared in *The Dominion*. Paley also cites Kohl and Farthing's *Impasse in Bolivia*, which notes that a no-longer available CIDA report from 2004 refers to an 800% return to Canadian businesses in the hydrocarbon sector.

12   Very little substantive information is actually on the project web site at: http://www.boliviacanadahydrocarbon.com/partners.htm. An evaluation conducted by CIDA (2008b) on its programming in Bolivia refers to CIDA's activities in the hydrocarbon sector as constituting the agencies 'flagship achievement' but offers little qualitative analysis of the actual program. IBM's website notes that it 'offers a range of services in the areas of exploration and production, refining and marketing; and provides advice on all aspects of industry restructuring, including regulation, liberalisation, private sector involvement and privatisation options.'

Given the strong centrifugal tendencies in Bolivia and the regional basis of political opposition, support to municipal associations may be susceptible to political manipulation. Yet the municipal associations are characterized by conflicting political tendencies and the question that must be asked in evaluating the contribution of Canadian actors is whether they have supported one political tendency at the expense of another. In this regard, FCM programme administrators seem to be sensitive to the local political dynamics and have, according to a programme manager for Bolivia, sought to 'foster a spirit of non-partisanship to reinforce a national voice for municipalities in general'.[13] FCM members have worked with regional associations of municipalities from both La Paz and Santa Cruz, though most programme activities are carried out by FCM itself in collaboration with FAM. FCM has also organized three exchanges with FAM representatives in Canada, ensuring that those delegates who participated were representative of the various regional tendencies.

One of the delegates who visited Canada, a councillor from El Alto and member of ACOBOL, notes that the women's association is characterized by similar political friction.[14] While ACOBOL may have emerged from within the liberal 'gender technocracy' that developed during the mid-1980s in contrast to more authentic expressions of the women's movement (Monasterios 2007), it has served as an important vehicle for promoting the inclusion of women in the political sphere and has been the driving force behind important initiatives, such as a law against gender-related political violence. While ACOBOL is undoubtedly divided by ethnic and class tensions, it remains a site of struggle in which indigenous women leaders are increasingly represented. However, the support that it has received through FCM and Rights and Democracy, as well as other donors, has largely served to reinforce the institutional capacity of the organization. As such, there is little evidence that one particular tendency has been supported at the expense of another.

Canadian democracy promotion has thus largely been defined by continuity in supporting state institutions regardless of who is in power. Although CIDA certainly sought to stabilize neoliberal polyarchy as the state increasingly came under attack, it did not strategically reorient its programmes to mobilize elite social forces against the growing power of the MAS and the indigenous movement. Nor did its support to polyarchy necessarily undermine other visions of democracy – as we shall see later, Rights and Democracy and Development and Peace supported grassroots organizations rooted in popular constituencies. The US pattern of engagement has been quite different.

---

13   Interview, Ottawa: 6 October 2009.

14   Interview, La Paz: 11 February 2009. ACOBOL has received funding from a number of democracy assistance agencies, including the NED and CIDA.

Responding to the MAS

The US Switch to Regime Change

Ultimately, US and Canadian interventions succeed even less at stabilizing the political situation than they did in Peru. In September and October 2003, large-scale uprisings during the gas war toppled the government of Sánchez de Lozada; the El Alto water war erupted in January 2005 and was followed by a second gas war in May and June that forced the resignation of the government of Carlos Mesa. Both governments failed to articulate a hegemonic order that would move beyond minor political-institutional concessions to include the substantive material and cultural demands being made by left-indigenous social forces. The state's project of passive revolution failed. As one political organizer in El Alto stated, 'These programmes are trying to establish Western structures of power, but transgressing is a part of the Aymara culture. They can offer all the courses they want. If we don't have water, jobs, sewage system, transportation, then when it's time to mobilize, we mobilize' (quoted in Lindsay 2005: 11).

Five months after the toppling of Goni's government, the United States supplemented its soft tactics with a more aggressive programme launched by USAID's Office of Transition Initiatives (OTI). The cornerstone of the new approach was a switch from support for the municipalities to support for the regional prefects, where right-wing forces throughout the *media luna* were preparing to resist the project of social transformation symbolized by the rise of the MAS.

The OTI programme launched in March 2004 initially complemented soft tactics aimed at managing social conflict by supporting moderate NGOs. Designed to 'reduce tensions in areas prone to social conflict and to assist the country in preparing for key electoral events', the programme disbursed US$13 million and more than 100 grants to departmental governments and civil society organizations through the US-based development firm, Casals & Associates (USAID 2007b). The programme initially focused on the city of El Alto, where the gas war had ended Goni's presidency, through community-based activities aimed at reducing conflict in the altiplano. However, after the December 2005 elections, OTI retargeted its programme toward 'building the capacity of prefect-led departmental governments' US democracy assistance for 2008 did not even mention the Bolivian government as a partner, and the focus moved to supporting NGOs, the private sector, and non-executive-branch entities of the government to combat the 'erosion of democracy' (Wolff 2011). This new phase of democracy promotion combined soft and hard tactics to contain the MAS by mobilizing right-wing prefects against it and by penetrating civil society on a deeper level to substitute radical demands for social change with the reformist approach of mainstream NGOs. With the rise of right-wing autonomist forces during this phase, US programmes were more successful in fostering destabilization than they had been in the previous phase in encouraging stabilization. By focusing on illiberal elements of political and civil society in the eastern lowlands, they also had a much more authoritarian flavour than the traditional focus on supporting polyarchy.

The initial wave of OTI grants went to infrastructural programmes in El Alto in an effort to depoliticize demands for greater control over resources and to demonstrate that the government could deliver basic services. From March 2004 to July 2005, $2.8 million dollars (nearly 43% of the money distributed by OTI) went toward 'information diffusion and dialogue' mainly in support of government public relations efforts such as those in support of a July 2004 national referendum on natural gas. According to Sandra Aliaga, head of communications under President Mesa until the end of 2004, OTI funded 40% of the government's public relations budget. The referendum avoided the issue of nationalization, and recipients of OTI grants who hosted workshops on the topic were instructed to focus on technical issues (Lindsay 2005).

These efforts were complemented by a new strategic focus on building a moderate indigenous movement as a counterweight in the east. The Brecha Foundation, an NGO founded by moderate leaders who emerged from the *Confederación de Pueblos Indígenas de Bolivia* (Confederation of Indigenous Peoples of Bolivia – CIDOB) with a reputation for being close to the departmental government in Santa Cruz, was selected to play this role. The NGO received unusually large grants of US$91,800 and US$67,600 to train indigenous leaders to participate in the constituent assembly first announced by the Mesa government. In February 2005 the OTI approved an additional grant of US$151,000 to Brecha to train 60 indigenous leaders from eastern Bolivia. According to director of Brecha, Víctor Hugo Vela, the NGO sought to groom indigenous leaders for participation in the assembly. Vela, who boasted of having met with Bush's Assistant Secretary of State for Western Hemisphere Affairs, Roger Noriega, during his visit to Bolivia in August 2004, openly opposed the MAS and called for getting rid of party leaders from indigenous associations (Lindsay 2005).

After the December 2005 elections, the OTI sought to undermine the growing hegemony of the MAS in the East by building the capacity of the departmental governments (Allard and Golinger 2009, Lindsay 2005). Technical administrative assistance and support for local job-creation schemes reinforced the ability of the prefectures to pose as ideological alternatives to the central state. USAID worked closely in particular with the regional prefects of Santa Cruz and Cochabamba (USAID 2006b), both of whom were key figures in the opposition to Morales. As Bigwood (2008) and Allard and Golinger (2009) point out, the OTI's work explicitly promoted 'sub-national de-concentrated' models of government in resource-rich departments that were preparing to launch autonomy referenda against the opposition of the central government. The referenda opened a new front in the reaction against the MAS by providing a regional alternative to the new vision of the nation being debated by the constituent assembly.

Apologists for the decentralization programme argue that it provided support to the prefects of all nine departments, including the three departments aligned with the MAS at the time (Chuquisaca, Oruro, and Potosí). A USAID official in La Paz defended the agency's programmes in an interview by noting that the MAS prefect of Oruro, Alberto Luis Aguilar, has been particularly supportive of the OTI's activities in his department.[15] But to accept this defence is to ignore the mountain of evidence of US hostility to the MAS and the obvious benefit of supporting the units of government that were its most ardent opponents. The programme may have supported all departments, but this could hardly have been otherwise without dragging the United States into another controversy during a period of international backlash.

Having benefited from US support, the prefects of Beni, Pando, Santa Cruz, and Tarija sought to counter the MAS-led project for constitutional change through referenda on regional autonomy, all of which passed in May and June 2008. When the regional governments moved to implement the autonomy statutes, however, the central government organized a recall referendum against eight of the nine regional prefects and the president and vice president. The conflict between the central government and the departments came to a head in August and September 2008, when autonomist forces attacked central-state institutions and indigenous peoples throughout the *media luna* in what Morales termed a 'civic coup'. In Pando, 20 peasants were killed by death squads supporting autonomy (Hylton 2008). These events revealed the close ties between the respectable face of the autonomist movement – the departmental prefects and the civic committees – and proto-fascist youth brigades such as the *Unión Juvenil Cruceñista* (Cruceñan Youth Union), which committed violent racist acts (Webber 2008).

As the crisis unfolded, US Ambassador Philip Goldberg was asked to leave the country for supporting the opposition. Goldberg, who had been nominated by President Bush and assumed his position in October 2006, was the former chief of the mission to Kosovo, where he had acquired experience in fanning regional tensions through support for separatist forces in that country (Allard and Golinger 2009). Documents leaked from the US embassy revealed Goldberg's desire to cultivate relations with indigenous groups opposed to the MAS in Chapare and the Media Luna soon after arriving in La Paz (Bigwood 2008). The embassy's diplomatic mission was called into question when a Fulbright scholar in Bolivia announced that he had been asked to spy on Venezuelans and Cubans. Similar allegations were made by Peace Corps volunteers. In the context of the autonomist attacks, Goldberg flew to Santa Cruz to meet with Governor Rubén Costa, a key leader in the autonomy movement. Immediately after the meeting, Costas ordered an 'official' takeover of national government offices in the region, suggesting that Goldberg had signaled his approval of such an action.

Goldberg's expulsion followed on the heels of Morales's decision to back the demands of the coca growers to expel USAID from the Chapare region because of its anti-coca programmes. The United States reciprocated by expelling the Bolivian ambassador and ending Bolivia's duty-free access to US markets under the Andean Trade Preferences Act. In November, Morales announced the suspension

---

15   Interview, La Paz: 17 February 2009.

of the operations of the DEA. In September 2009 Morales accused USAID of conspiring against him by providing funding to the opposition presidential and vice presidential candidates. In July 2010 the mayors of Pando went so far as to expel the agency from their territory, and Morales threatened to do the same at the national level.

The Combined Approach

The hard tactics of US democracy promotion during this period were to some extent obfuscated by ongoing party-strengthening initiatives and support for dozens of moderate NGOs whose relationship to the MAS was more ambiguous. With the inauguration of the constituent assembly in 2006, one moderate coalition of 41 NGOs, the *Red de Participación y Justicia* (Network for Participation and Justice – PJ), began receiving a steady stream of funding from USAID and the NED. PJ was the main implementing organization of a US$7.5 million project entitled Support for Civil Society and Justice Reform, which ran from 2002 to 2008. During the constituent assembly, it organized workshops and forums with various sectors of civil society, reaching more than 15,000 Bolivians. Despite its support for communal justice, PJ is by no means a radical coalition. Its approach to justice and democracy has focused on anticorruption campaigns, state transparency, and accountability, issues that help frame the democratic discourse in the public sphere away from structural antagonisms and undercut support for anti-systemic forces. USAID has explicitly encouraged this focus by inviting more successful anticorruption coalitions in the region to provide PJ with training (Berthin et al. 2005), attempting to reproduce the Peruvian model discussed in the previous chapter.

CIDA has also provided modest support to Bolivia's anti-corruption network through a project that ran from 2007 to 2011 with a total budget of $800,000, implemented by the Chilean-based NGO, *Corporación Participa* (Participatory Community) (CIDA 2010c). Through CIDA and NED funds, the Chilean NGO has sought to create a regional network of civil society organizations dedicated to promoting anti-corruption issues. Although the amount of Canadian assistance that supports *Participa*'s work in Bolivia is minimal,[16] CIDA's role in creating this regional coalition illustrates how Canada has increasingly sought to frame democracy in terms that are congruous with its economic interests.

At the same time, a report by the PJ on the 2006 presidential elections commended Morales and his vice president, Álvaro García Linera, for their conduct and overwhelming support (PJ 2006). The coalition and many of its member organizations have also signed statements denouncing the actions of the regional opposition. One of its members, the *Casa de la Mujer* (House of Women

---

16   CIDA's disclosure of funding notes that only 4.5% of project funds cover activities carried out in Bolivia (CIDA 2010c). The same amount is also allocated to activities carried out in Peru.

– CM), based in Santa Cruz, has allied itself with other NGOs in denouncing racist acts of violence against popular organizations in Santa Cruz (see Mokrani and Uriona 2008). But the moderate political orientation of the coalition and many of its members should not obscure the fact that many have been cultivated by US agencies to counterbalance the hegemony of the MAS in indigenous civil society. The president of the coalition, Eduardo Barrios Sánchez, is a founding member of the *Instituto Politécnico Tomás Katari* (Tomás Katari Polytechnic Institute – IPTK), one of the largest rural development NGOs and one that was highly critical of the role of the MAS in the constituent assembly (see Barrios 2007). The driving force behind the organization, Franz Barrios Villegas, was president of a centre-left party that joined Lozada's political coalition during his second administration (though his party, the *Movimiento Bolivia Libre* (Free Bolivia Movement – MBL), officially supported the MAS in the December 2009 elections (FM Bolivia 2009)).

Another member of the coalition with an anti-MAS orientation, the *Instituto de Investigación y Capacitación Pedagógica y Social* (Institute of Educational and Social Research and Capacity Building – IIPS), has received a regular stream of grant funding from the NED since the mid-2000s. According to one representative who was interviewed, the organization emerged from the ranks of the radical teachers' union movement in the 1990s, which fought alongside indigenous parties, including the MAS.[17] The representative argued that US organizations that have received funding from the NED are by no means expected to advance US interests, but an IIPS funding request tells a different story, denouncing the 'anti-democratic, radical opposition' and warning that 'in the face of a lack of leaders who truly represent them, radical leaders such as Evo Morales and Felipe Quispe have risen to prominence' (Bigwood 2008). The IIPS has refrained from publicly denouncing the MAS, however, most likely because of the party's near-hegemonic position. Its dual discourse suggests that indigenous organizations that position themselves as moderate alternatives have to be careful about alienating their support base.

Soft tactics have also undermined the MAS without openly mobilizing the elite opposition through support for moderate NGOs led by urban middle-class professionals critical of the constitutional project. One organization receiving US funding that criticized the constitutional process was the *Fundación de Apoyo al Parlamento y a la Participación Ciudadana* (Foundation for Parliamentary Support and Citizen Participation – FUNDAPPAC), an NGO made up of former parliamentarians. During the process, it hosted public meetings and published a regular bulletin on the latest developments surrounding the assembly with NED funding. Although the bulletins presented diverse perspectives, including those put forward by MAS representatives, the NGO's report argued that the process violated numerous laws and that the proposed constitution failed to create a new social pact responding to the aspirations of all Bolivians (Rivera 2008). The report

---

17   Interview, 10 February 2009.

also lamented the exclusion from the process of organizations representing the interests of the urban-based middle class and the departmental civic committees.

The *Fundación Boliviana para la Democracia Multipartidaria* (Bolivian Foundation for Multiparty Democracy – FBDM) took a similar position on the constitution. The FBDM, an organization dedicated to political pluralism, was an implementing partner of the USAID-funded Support for Democratic Processes along with IRI and NDI. One of its representatives decried the polarization of the country and noted that the organization sought to bridge the divide between different political actors.[18] Among other things, it hosted meetings between different parties and civil society organizations, publishing numerous booklets on the constitutional project as well as a lengthy critique of the proposed constitution (FBDM 2007). Another NED-funded NGO, former President Mesa's *Fundación Comunidad* (Community Foundation), took a high-profile public stance against the constitution, arguing that it broke with the basic principle of equality.[19] Although not all NGOs that received US funding during the constitutional process opposed the MAS, many did. While the ethnocentric assumptions of the democracy promotion industry typically forbid such comparisons, one can imagine the sense of scandal and outcry in the United States had the MAS (or, worse, the government of Hugo Chávez) contributed funds to civil society groups engaged in the national debate on health care reform.

Canadian Democracy Assistance and Popular Forces

Although Canadian democracy assistance to Bolivian civil society is generally quite modest, both Development and Peace and Rights and Democracy have provided support to various civil society organizations, including progressive NGOs and social movements with a mass base. Amongst these movements figure some of the most important indigenous and popular organizations, most of whom supported the proposed constitution – albeit not without criticism of some of the actions MAS representatives in the Constituent Assembly.

In examining Canadian democracy assistance to Bolivian civil society, the most obvious trend is the near absence of CIDA aid to civil society organizations. Throughout most of the 2000s, CIDA assistance has been channelled primarily to state institutions. In 2007–2008, CIDA (2009) began providing significant support to women's organizations, though all 10 grants that were awarded that year went to organizations working primarily on health and reproduction issues.

Development and Peace's Bolivia Country Program contributed $923,724 to Bolivian civil society organizations in its 2003–2006 phase, and has allocated an additional $2.56 million for its five-year programme from 2006 to 2011 (Development and Peace 2006a, 2006b). Initially, the programme provided support to 13 grassroots organizations, though this was subsequently reduced to six in

---

18   Interview, 13 February 2009.
19   Interview, 18 February 2009.

its second phase. The main objective of the programme has been to strengthen the capacity of social actors to promote change. Some of its partners, most notably the national federation of domestic workers, the *Federación Nacional de Trabajadoras del Hogar de Bolivia* (FENATRAHOB), and the regional association of indigenous peoples in Santa Cruz, the *Coordinadora de Pueblos Etnicos de Santa Cruz* (CPESC), have been at the forefront of popular struggles.

Founded in the late 1980s, FENATRAHOB represents the domestic workers of Bolivia, the vast majority of whom are indigenous women who have migrated from the countryside to work in the urban homes of the wealthy and middle class. One of its founders and leaders, Casimira Rodríguez, served as Morales' first minister of justice in 2006, during which time the government passed a Declaration of the Rights of Domestic Workers (Romer 2008). FENATRAHOB's leaders situate their struggle to counter patriarchal racist relations between domestic workers and their employers within the overall struggle against the legacy of colonialism (Blofield 2009).

CPESC is the regional affiliate representing the indigenous of Santa Cruz within the national indigenous federation, CIDOB. Formed in 1992, Van Cott (2005) characterizes the association as a social movement organization which groups together multiple base community organizations. CPESC has maintained close ties with the MAS and was a key supporter of the constitutional project. Indeed, the organization helped galvanize the process through the March for the Constituent Assembly and Natural Resources in 2002, in which members marched for 37 days from Santa Cruz to La Paz. In September 2008, its headquarters were looted and nearly destroyed, an act it alleged was committed by vandals associated with the Prefect, the right-wing UJC, and the Comité Cívico Pro Santa Cruz (Civic Committee of Santa Cruz – CCPSC) (CPESC 2008).

Development and Peace has also supported progressive NGOs which have acted in solidarity with popular movements, such as the organization *Alas Yvi Avarenda* (Land of the People). Although *Alas Yvi Avarenda* does not possess a broad social base like FENATRAHOB and CPESC, it carries out its work in solidarity with popular organizations in Santa Cruz. Established in 1991, the NGO focuses on building the capacities of popular organizations engaged in struggles surrounding land claims, control of natural resources and the protection of the environment. The organization has signed numerous declarations in support of popular causes, including land reform. It has also supported the *Coordinadora Regional por el Cambio* (Regional Coordinator for Change – CORECAM), the body linking regional social movements, in denouncing the project of regional autonomy as one which represents the interests of the regional elite rather than indigenous workers and peasants. Although Development and Peace did not continue supporting CPESC in its second phase of programming, it maintained its support to FENATRAHOB and *Alas Yvi Avarenda*, as well as other progressive NGOs.

Rights and Democracy only established its Bolivia country programme, Strengthening the Participation and Capacity of Indigenous Organizations, in

2006.[20] Its main objective has been to foster greater political participation of indigenous peoples and women, particularly in the current process of social change. Through this programme, the Canadian QUANGO has provided support to CEADESC, which, as noted earlier, was one of the few US-funded organizations not to oppose the constitution. More significantly, however, Rights and Democracy has supported an alliance of indigenous, peasant, and women's organizations called the *Pacto de Unidad* (Unity Pact), all of which are linked to mass social constituencies. The most politically significant of these organizations is the *Confederación Sindical Única de Trabajadores Campesinos de Bolivia* (Union Confederation of Peasant Workers of Bolivia – CSUTCB), which was one of the first indigenous peasant organizations to emerge in the late 1970s. Another member, the *Federación Nacional de Mujeres Campesinas, Indígenas Originarias Bartolinas Sisa* (National Federation of Peasant Women, Indigenous Origins Bartolinas Sisa – FNMCIOB-BS), is the main popular organization bringing together peasant women. Although the *Pacto* was strongly in favour of constitutional change, it was at times critical of decisions made by MAS representatives in the Constituent Assembly that sidelined their proposals. The executive secretary of the CSUTCB, Isaac Ávalos, was particularly outspoken about the behaviour of certain MAS representatives, which he considered treacherous (La Razón 2006). Nonetheless, the *Pacto* did manage to ensure that most of its demands were incorporated in the proposed constitution, for which the CSUTCB campaigned strongly in favour (Bolpress 2008).

Canadian democracy assistance thus provided support to many popular organizations in Bolivia during a time of social mobilization and upheaval. Whether Development and Peace will continue to support these organizations, however, is an open question (as previously noted, Rights and Democracy, for its part, has been shut down). As discussed in Chapter 2, the new focus on funding Canadian NGOs in the context of the Andean Regional Initiative for Promoting Effective Corporate Social Responsibility has led to the further subordination of Canadian development assistance to the needs of Canadian capital. Development and Peace's opposition to the mining practices of Canadian-based multinationals is unlikely to endear it to the Harper government.

Concluding Remarks

This chapter has argued that the United States and Canada have both sought to influence Bolivia's process of democratic development, albeit from considerably different vantage points. For the United States, an initial phase of democracy

---

20   The founder and president of the aforementioned *Fundación Boliviana para la Democracia Multipartidaria*, Guido Riveros Franck, sat on the board of directors of Rights and Democracy.

promotion coincided with the state's attempt to contain the popular revolt of left indigenous social forces through a project of inclusive neoliberalism and passive revolution. Soft tactics in this context reinforced state institutions and moderate NGOs articulating liberal notions of citizenship. Hard tactics supplemented the US effort beginning in 2004, when the OTI began mobilizing opposition social forces, including moderate indigenous NGOs and, after the December 2005 elections, the departmental prefects opposed to the central government. This strategic shift, however, coexisted with large-scale support for moderate civil society organizations and NGOs whose relationship to the MAS was ambiguous, including some that did not oppose the central government and others that were in fact critical of the departmental prefects.

For Canada, a strategy of regime change did not develop during the period under consideration, though its democracy assistance programmes did seek to stabilize the neoliberal state as it increasingly came under attack in the first half of the decade. Canada also used its development programmes to influence the management of key natural resources. Nonetheless, it continued to support state institutions after the rise of the MAS while Canadian organizations provided assistance to progressive NGOs and social movements with mass social constituencies.

By way of conclusion, I offer three general remarks on the MAS's response to US democracy promotion to shed some light on this practice in the regional order in the new conjuncture and its implications for North American democracy promotion as a whole. First, the MAS's combination of resistance and accommodation to US democracy promotion indicates that US imperial pressures are significant despite the changing balance of regional power. In addition to shutting down DEA operations and the USAID presence in Pando and Chapare, the Morales government forced USAID to realign its programmes with government objectives by shifting its orientation from the departments to the municipalities and recognizing the central government as a key partner (Wolff 2011). At the same time, the United States still has considerable leverage over Bolivia, which lost around US$63 million in manufacturing exports after Bush suspended its duty-free access to the US market (Skeen 2010b). The importance of US assistance in areas such as health should also not be discounted (the total USAID programme budget for 2001–2009 was US$881.5 million, of which only US$100 million went to democracy assistance).[21] For the time being, the MAS seems willing to tolerate democracy promotion programmes as long as they are aligned with its own strategic objectives. The re-establishment of duty-free access to the US market – not to mention the ongoing importance of development assistance – provides a strong incentive for the MAS to avoid escalating the confrontation. When Bolivian Foreign Minister David Choquehuanca and US Assistant Secretary of State Arturo

---

21   I base these figures on data obtained from USAID congressional reports (2002a, 2003, 2004, 2005b, 2006a, 2007a, 2008b, 2009).

Valenzuela met in June 2010 to begin discussions on normalizing relations, Morales declared his own hope that the two countries would now 'advance with this new framework agreement for full diplomatic, trade and investment relations' (Main 2010). Soon afterward, Morales again accused USAID of supporting opposition groups but did not go so far as to expel the agency. At the time of writing, the two governments recently signed the new framework, which Bolivian Vice President García Linera refers to as a unique collaborative model in Latin American-US relations. Among other things, the framework stipulates that the DEA will not return to Bolivia, that the *cocalero* union will be involved in eradication efforts (as well as Brazil) and that USAID projects can be called to a joint-committee for review (Achtenberg 2011).

Second, US democracy promotion programmes have become more sophisticated. As we have seen, party-strengthening programmes in Bolivia since 2004 have included the MAS through multiparty initiatives. This marks a departure from earlier programmes in Bolivia and in Haiti. Even the OTI programme provided some benefits to MAS-led departments. But it is clear that the limited benefits derived by the MAS through its participation in specific programmes is overshadowed by the greater gains reaped by its political enemies. What is more, the pronouncements of US officials and the available documentation show beyond any doubt that US programmes have sought to undermine the MAS while containing left indigenous social forces. What seems likely is that, under the spotlight, the United States has sought to obfuscate its political objectives through soft tactics that provide a shield of deniability from charges of political manipulation. Multiparty training programmes and massive civil society support are likely strategic responses to the shifting balance of power and the need to retain credibility in the context of the regional and international backlash. With several Latin American countries now rejecting the traditional patterns of US interventionism, the extent to which democracy promotion tactics can be used to openly destabilize and undermine left and centre-left governments has been significantly constrained.

Conclusions

Neoliberal Hegemony without North American Leadership

This book has argued that North American democracy promotion cannot be abstracted from the foreign policy interests and political objectives that guide US and Canadian engagement in the Americas. At root, it has more to do with securing stable social orders in a context of neoliberal regionalism than it does with the lofty and idealistic aims evoked by its discourse. Since the 1990s, both Canada and the United States have linked the promotion of democracy to the construction of a regional hemispheric order based on the free mobility of capital, liberal-democratic institutions and, over the last decade, securitization to deal with the social fallout of persistent poverty and inequality. While the institutionalization of a liberal-democratic order in the hemisphere is not to be dismissed, democracy promotion itself has often served as a hegemonic practice reducing the scope of democracy to the requirements of neoliberal accumulation. As Latin American popular movements have struggled to deepen democratic participation and democratize the economy, democracy promoters have responded by containing and defusing threats to the neoliberal state.

This has occurred through strategies of regime change directed against enemy governments as we saw in Haiti and Bolivia, where democracy assistance was used to empower opposition forces against the governments of Jean-Bertrand Aristide and Evo Morales. It has also occurred through the stabilization of allied states like Peru and Bolivia before Morales' 2006 presidential victory, where democracy assistance contributed to the state's project of passive revolution by attempting – albeit unsuccessfully – to de-politicize conflicts that threatened the social order. For the United States, democracy promotion emerged in continuity with the tactics of intervention and counter-insurgency pioneered in an earlier, more authoritarian, phase of regional order. For Canada, strategies of stabilization and regime change represent the corollary of a lengthy process of neoliberal realignment, though one that has not yet fully eclipsed the legacy of a more progressive era of international development and its grassroots traditions, as we saw in Peru and Bolivia. As the case of Haiti illustrates, however, we should not romanticize this tradition either – democracy promotion is always inscribed in unequal power relations between North and South, and Canadian NGOs can align themselves with reactionary social forces depending upon local circumstances. Ultimately, the ability of all democracy promoters to affect local power relations without any accountability speaks to the democratic deficit at the very heart of democracy promotion.

Looking ahead, what can be said about the future of North American-led regional order – and the role of democracy promotion in reproducing it – given the important political changes that are sweeping the Americas? To conclude, I identify three important trends in the evolution of regional order and North American democracy promotion based on the evolving balance of power between classes and states, re-connecting observations made in the context of the case studies to the macro-developments discussed in Chapter 2. While it would be foolish to think that these trends point in an absolute or definitive direction, my comments are intended to contribute to and spark debate for a critical research agenda on North-South relations in the Americas. I end with a few words on how Canada and the United States have much to learn from the peoples of Latin America in defending against 'democratic backsliding' in their own countries.

Constraints and Continuity in a Multipolar Regional Order

First, the emergence of a multipolar hemisphere will further restrict the leadership role that the United States and Canada have assumed in articulating regional order. As neo-Gramscians have reminded us, hegemony in international relations is based not only upon military power but ideological resources to structure the choices and behaviour of competing and lesser powers in ways that favour the interests of the most powerful state, in particular its desire to remain the pre-eminent actor (Cox 1993). As Carranza (2010) argues, the ability of the United States to exercise hegemony in Latin America has been damaged by the unfulfilled promises of neoliberalism and by the Bush administration's unilateralist foreign policies in the 2000s. This has resulted in a disjuncture between the structural military-economic dimensions of US power and its 'agential' dimensions, or ability to lead by influence. Canada's attempts to leverage its status as a middlepower to legitimize *Pax Americana* have done little to rehabilitate North American leadership; the exclusion of both countries from the CELAC has institutionalized this failure.

Brazil's rejection of the FTAA was the first sign of this over a decade ago, but there are now additional counter-balances in the form of regional integration projects and the rise of China – for better or worse – as the region's fastest growing trading partner. Brazil has also pushed for South-South alliances with China, India, Russia and South Africa for changes in what they rightfully claim to be an unjust world economic order (though many in these countries do in fact benefit from it). This is not to fall prey to a sort of Global South or Pink Tide utopianism, where one's faith is put in the emergence of a more pluralistic, democratic and perhaps socialist order simply because there are left and centre-left governments which have played an important role in calling for increased South-South cooperation, and Washington's influence is on the decline. Latin American states will continue to be divided between left and right, with a resurgent right in countries where the left has done little to dismantle the structural power of the agro-mineral financial oligarchy. As Mace, Cooper and Shaw (2011) argue, moreover, international and hemispheric relations will continue to be defined by a hierarchy of powers as South-South forums such as the G20 privilege some states in contributing to global governance over others. As the new forms of governance blend traditional powers with new ones from the Global South such as the BRICSs, peripheral states in Central America and the Caribbean will largely remain excluded. The decline of US influence cannot be overstated either: 'not many people think like that', Russell (2011) reminds us, 'in Colombia, Mexico, Central America, or the Caribbean'.

But multipolarity and the new regional counter-balances to US domination suggest that blatant interventionism of the kind witnessed in Haiti and Bolivia on the part of the United States or Canada will no longer be considered acceptable. On the plus side, this will provide some additional space for radical governments in control of the state to contest the power of dominant groups and classes – though stabilization tactics in allied states will most certainly seek to prevent radical parties from winning political power in the first place. Multipolarity has also lead to regional forms to denounce US interventionism – such as the ALBA, where democracy promotion has been regularly criticized. More generally, as the work of Robert Cox (2002, 2005) has emphasized, the hopes for a better world lie in the reconstruction of an effective state system as a counter-weight to US dominance and its long-term trajectory of decline, and multipolarity in Latin America at least opens this possibility.

As for regional relations and the political economy of regional order, the cleavage between those states seeking to increase consensus and those relying primarily on coercion is likely to remain. The internationalization of the state has deeply entrenched neoliberalism, and the ongoing process of de-industrialization will reinforce the class power of the export-oriented agro-mineral oligarchy across the Americas. Both those states which have instituted a more socially-oriented form of neoliberalism and those which have managed the contradictions of neoliberalism through repression will continue to face discontent. As Robinson (2004) argues, 'it is not clear how effective national alternatives can be in transforming social structures, given the ability of transnational capital to utilize its structural power to impose its project even over states that are captured by forces averse to that project'. The deepening of democracy in the hemisphere will depend upon the ability of the social movements to combine political vision with leadership in a project to transform the state. What remains to be seen is whether the more radical regional bloc organized under the ALBA will crystallize into coherent alternative with its own distinct regional political economy. There are few signs of this happening right now, but it should remain a tactical focus of the Latin American left since it is the only regional project with the potential to move beyond neoliberalism.

Soft Tactics and Stabilization

Second, material and ideological assistance from both Canada and the United States will continue to provide regional allies with resources to support their hegemonic projects, despite the overall decline in North American influence and leadership. On the consensus side of the equation, soft tactics linked to democracy promotion in support of inclusive neoliberalism such as those that were examined in Peru and in Bolivia in the early 2000s will likely constitute the new norm. Those governments unable or unwilling to create an integral hegemonic order through class compromise will welcome US and Canadian democracy assistance programs that contribute to passive revolution. Ironically, this may include parties, movements and leaders once deemed anti-systemic such as Ollanta Humala, whose record thus far has been perfectly in keeping with the inclusive neoliberal paradigm. In this respect, the ideological vision and limited reforms promulgated by North American democracy promotion are very much in line with the hegemonic aspirations of many states. Thus, declining leadership will not necessarily translate into declining force of vision. On the coercion side, security assistance and the War on Drugs will continue to funnel massive amounts of aid to security-state polyarchies containing the fallout of class inequality and neoliberalism through force.

Looking forward, Canada and the United States will likely use democracy assistance programs to impose a controlled democratic transition on Cuba should the Castro government move further along the path of market liberalization. Here, Canadian efforts could be particularly instrumental given the stronger bilateral relations between the two countries. We might also expect to see Canada play an expanded role in stabilizing the Caribbean more broadly given its historic ties to the region and the Harper government's own ambitions in turning Canada into a regional power. The enhanced presence of the Canadian military in the region certainly points in that direction. Undoubtedly, the United States will continue to focus its democracy assistance efforts on Colombia and Mexico given their geopolitical and strategic importance. How these efforts reinforce the policies of coercion in the making of hegemony is a topic that requires further comparative research.

Additionally, the role of the United States and Canada in supporting the hegemonic aspirations of allied states despite the overall decline of North American influence in the hemisphere raises questions on the nature of hegemony in both world and regional order. Neo-Gramscians have traditionally associated hegemony with leadership, and hegemonic order with the dominance of a powerful state or states able to impose a project which incorporates the interests of subordinate states and social forces. Neoliberalism in Latin America has never been truly hegemonic in that sense, but it did achieve the support of the majority of Latin American governments under US leadership. The ongoing primacy of the global economy in configuring the regional regime of accumulation coupled with the reassertion of the state in mediating between the global market and society suggests that hegemonic power is now more diffuse; neoliberalism remains deeply entrenched even as US power wanes. This suggests the need for deeper empirical studies on the transnational linkages between dominant classes across the hemisphere and the role of the emerging regional powers in expanding and mediating opportunities for accumulation. To what extent can we speak of regional class formation along hemispheric lines? What is the relationship of dominant classes across the region to the United States? Are there new tensions and contradictions between open regionalism and global capitalism as the balance of power between states shifts? To what extent will new regional powers such as Brazil take on the disciplinary function once monopolized by Washington?

Ongoing North American Convergence

The third and final point is that Canadian and US approaches to democracy promotion will likely undergo further convergence. As we saw in Chapter 2, both countries have increasingly focused democracy assistance on strategic allies facing widespread discontent and social fallout from neoliberal policies. But where the United States has used democracy assistance to build the legitimacy of governments, the Canadian approach to stabilization has focused more on strengthening state institutions and the political system as a whole. These differences were observed in some depth in both Peru and Bolivia.

With the development of a more compliant network of NGOs, however, Canadian stabilization efforts are likely to take on a more ideological character in the ongoing democracy wars across the hemisphere. We can expect a diminishing role for the grassroots approach and a more explicitly political focus on the part of Canadian state agencies and a new breed of NGOs. True, the argument can be made that the grassroots approach was already dead when Canadian NGOs adopted a strategy of regime change in Haiti in keeping with the approach of IRI and USAID. As I argued in Chapter 3, however, this overlooks the historical ties that Canadian NGOs had developed with their Haitian counterparts that led them to internalize the political position of elite civil society. The implication is that the grassroots tradition was still relevant, an argument that was substantiated empirically when we examined the types of organizations that were supported in Bolivia and Peru, many of which were firmly rooted in popular civil society. Seen in this light, the actions of the Harper government take on added weight, and new initiatives such as the expansion of the democracy pillar under the Americas Strategy must be regarded with a great degree of suspicion. Together, the many changes could amount to a significant institutional and cultural shift in the development field of practice and its ultimate subordination to the state. With these changes well underway, there is a need to further investigate convergence in other policy areas like security-defence, diplomacy, aid, and commercial relations to better understand how Canadian and US imperial practices complement each other and vary in these domains in a time of great regional flux. These are important lines of inquiry in the development of a critical regionalism of North-South relations in the Americas.

For those who imagine a more constructive role for Canada and the United States in the world and recognize the importance of combining critical theory with alternative policy thinking, it is tempting to end by asking the question – is there a role at all for either state to play in supporting democracy abroad? If so, what kind of a role is it? Certainly, one must consider what more progressive North-South relations would look like in addition to critiquing those that exist. In a less hierarchical regional order where Canadian and US foreign policy were not so fully subordinated to economic interest, perhaps such a role could be more easily defined. But until the foreign policies of both countries have undergone a major reorientation, engaging such questions assumes that they have the disinterest and moral standing to be providing others with advice on how they should be governing themselves. For the time being, the credibility gap between self-interested motives and the requirements for genuine democratic development is simply too large.

What seems obvious right now is that Canada and the United States have a lot to learn from the peoples of the Americas in their struggles to deepen democracy. Compared to the state of democratic discourse in North America, many Latin American countries are alive with foundational questions that speak to the very definition of democracy today. Indeed, the popular mobilizations across the Americas stand in sharp contrast with the moribund liberal democratic system of the United States – and to a lesser extent Canada – where the divide between the promise of liberal democracy and its actual practice is growing each day. The contrast exposes the shallowness of the implicit cultural claims to superiority that justify democracy promotion as a legitimate enterprise. Whatever form of democracy is emerging in many Latin American countries, it is clear that it is not the place of Canada or the United States to shape this process. The monopolization of political power by a plutocracy in the United States, the outrageous barriers to participation in political life imposed by campaign finance laws and the inability of the party system to channel the interests of the poor and excluded suggest that the focus on democratic backsliding in the hemisphere could just as easily be reoriented northward. In Canada, too, democratic backsliding is reaching an alarming pace as executive power is further concentrated under the office of the prime minister, parliamentary debate is circumvented or shut down, independent agencies have come under attack, and the governing party engages in voter suppression. In both countries, declining voter turnout – particularly amongst youth – further jeopardizes the legitimacy of the democratic system. Although there are some glimmerings that a more critical consciousness could be in the making – for example, in the new awareness of growing inequality thanks in no small part to the Occupy Wall Street movement – there is much work to be done to reverse and prevent the further degeneration of democracy. In this respect, North Americans have many lessons to learn from the experiences of Latin American popular movements, which have been at the forefront of resistance to the exclusion and dehumanization imposed by the neoliberal state.

Bibliography

Achtenberg, E. 2011. A political victory for Bolivia. *NACLA Report on the Americas* [Online, 19 November]. Available at: http://nacla.org/blog/2011/11/19/political-victory-bolivia [accessed: 2 January 2012].

ADEHRPERU. 2007. *CIDH pide reabrir investigación de "El Frontón"* [Online]. No longer available [accessed: 8 January 2008].

Alarcón, C. 2009. *Entre Sodoma y Gomorra* [Online: Fundación Comunidad]. Available at: http://www.comunidad.bo/prensa/desarrollo_nota.shtml?x=357 [accessed: 15 December 2009].

Alban Guevara, R. 2005. *The Role of the Private Sector in Rebuilding Haiti* [Online: Inter-American Dialogue]. Available at: http://www.focal.ca/pdf/Haiti_Role%20Private%20Sector_Brief_October%202005.pdf [accessed: 26 November 2008].

Albó, X. 2002. From Indian and campesino leaders to councillors and parliamentary deputies, in *Multiculturalism in Latin America: Indigenous Rights, Diversity, and Democracy*, edited by Rachel Siede. Houndmills: Palgrave Macmillan, 74–102.

Aldunate, E. 2010. *Backpacks Full of Hope*. Waterloo: The Centre for International Governance Innovation and Wilfred Laurier University Press.

Allard, J.G. and E. Golinger. 2009. *La Agresión Permanente*. Caracas: Ministerio del Poder Poder Popular para la Comunicación y la Información.

Allen, M. 2005. The politics of democracy promotion. *Democratiya* [Online, 1, Summer]. Available at: http://dissentmagazine.org/democratiya/article_pdfs/d1Allen.pdf [accessed: 20 October 2008].

Alonso Ramos, A. 2008. *Investigarían a ONGs incómodas* [Online: Adehr Peru]. Available at: http://www.adehrperu.org/noticias-ddhh/investigar-an-a-ongs-inc-modas.html [accessed: 28 July 2009].

Alvarez, S., E. Dagnino and A. Escobar. 1998. *Cultures of Politics Politics of Cultures: Re-Visioning Latin American Social Movements*. Boulder: Westview Press.

Americas Policy Group. 2009. *What Role for Canada in the Americas: Statement of the Americas Policy Group* [Online, 15 April]. Available at: http://www.ccic.ca/_files/en/working_groups/003_apg_2009-04-15_statement_what_role_for_cda_in_americas.pdf [accessed: 2 May 2009].

Amnesty International. 2005. *Arbitrary Arrest/ Prisoner of Conscience: Haiti, Gérard Jean-Juste (m), aged 59, Catholic Priest* [Online: AI Index: AMR 36/008/2005, 25 July]. Available at: http://ijdh.org/archives/15981 [accessed: 10 June 2009].

Anderson, P. 2011. Lula's Brazil. *London Review of Books*, 33(7), 3–12.

Antrobus, P. 2004. *The Global Women's Movement: Origins, Issues and Strategies*. Dhaka: the University Press.

Arce, M. 2008. The repoliticization of collective action after neoliberalism in Peru. *Latin American Politics and Society*, 50 (3), 37–62.

ARD inc. 2003. *Report on Organized Policy Issues Dialogue between Haitian Civil Society Organizations and National Public Institutions* [Online: USAID report]. Available at: http://pdf.usaid.gov/pdf_docs/PNACX904.pdf [accessed: 1 June 2009].

Arkonada, K. 2009. *Vivir Bien: El Paradigma Indígena en Debate* [Online: CEADESC]. Available at: http://www.ceadesc.org/2009/10/vivir-bien-el-paradigma-indigena-en-debate/ [accessed: 22 August 2009].

Arnold, R. 2012. *Peruvians Oppose CIDA's Joint CSR Initiative with Barrick Gold and World Vision* [Online: Mining Watch Canada, 9 March]. Available at: http://www.miningwatch.ca/article/peruvians-oppose-cida-s-joint-csr-initiative-barrick-gold-and-world-vision [accessed: 20 March 2012].

Asamblea de la Red LAC. 2008. *Primera Asamblea de la Red Latinoamericana y del Caribe para la Democracia* [Online]. Available at: http://www.democracialatinoamerica.org/uploads/I_Asamblea.pdf [accessed: July 6, 2009].

Associated Press. 2005. Rumsfeld's tour of South America is directed at promoting stability. *New York Times* [Online, 18 August]. Available at: http://www.nytimes.com/2005/08/18/international/americas/18rumsfeld.html?_r=1&scp=1&sq=rumsfeld+toledo&st=nyt [accessed: 24 February 2008].

Axworthy, T., L. Campbell and D. Donovan. 2005. *The Democracy Canada Institute: A Blueprint*. IRPP Working Paper Series, no. 2005-02a.

Barr, R. 2003. The persistence of neopopulism in Peru? From Fujimori to Toledo. *Third World Quarterly*, 24 (6), 1161–1178.

Barrig, M. 2002. La persistencia de la memoria: feminismo y estado en el Perú de la década de 1990, in *Sociedad Civil, Esfera Pública y Democratización en América Latina: Andes y Cono Sur* edited by Aldo Panfichi. Lima: Universidad Católica del Perú, Fondo de Cultura Económica, 578–609.

Barrios Villegas, F. 2007. *El Cambio en Bolivia* [Online: Punto de Vista, 1–4 September]. Available at http://www.sodepaz.org/america-latina-mainmenu-15/bolivia-mainmenu-22/240-el-cambio-en-bolivia.html. [accessed: 9 July 2009].

Barry, T. 2005. *World Movement for Democracy--Made in the USA* [Online: Foreign Policy in Focus, July 29]. Available at: http://www.fpif.org/fpiftxt/175. [accessed: 10 January 2010].

Barry, T. 2007. *The New Politics of Political Aid in Venezuela* [Online: Americas Program, Center for International Policy Report, 24 July]. Available at: http://www.globalresearch.ca/index.php?context=vaandaid=6391 [accessed: 8 January 2010].

Batay Ouvriye. 2003. *Déclaration de Batay Ouvriye sur la Situation du Pays Aujourd'Hui* [Online]. Available at: http://www.batayouvriye.org/downloads/ithpfr.pdf [accessed: June 18, 2009].

Bazo, F. 2006. *Congressional Candidates Challenging Discrimination in Peru* [Online: Peru Election 2006: The University of British Columbia, 3 March]. Available at: http://weblogs.elearning.ubc.ca/peru/archives/023609.php [accessed: December 2008].

Bebbington, A., M. Scurrah and C. Bielich. 2008. *Mapping Current Peruvian Social Movements*. Manchester: Peruvian Centre for Social Studies.

Becker, D. 1983. *The New Bourgeoisie and the Limits of Dependency: Mining, Class, and Power in "Revolutionary" Peru*. Princeton: Princeton University Press.

Bedrossian, T. 2012. No plans to release Americas strategy consultation. *The Embassy*, 394, March 14.

Beeton, D. 2009. *Is the United States Funding Violent Opposition Groups in Bolivia?* [Online: The Democracy Center, 6 May]. Available at: http://democracyctr.org/blog/2009/05/is-united-states-funding-violent.html [accessed: 8 August 2009].

Berthiaume, L. 2010a. Kent blasts Venezuela's Chavez following visit: But the minister of state for the Americas has high hopes for Bolivia. *Embassy*, January 27.

Berthiaume, L. 2010b. Leading the change at Rights and Democracy. *Embassy*, September 22.

Berthin, G. et al. 2005. *Anti-Corruption and Transparency Coalitions: Lessons from Peru, Paraguay, El Salvador and Bolivia* [Online: USAID report, August]. Available at: http://info.worldbank.org/etools/ANTIC/docs/Resources/Country%20Profiles/Paraguay/USAID_ACTransparencyCoalitons.pdf [accessed: 5 July 2009].

Bigwood, J. 2006. U.S. meddling in Peruvian presidential race? *Znet* [Online, 20 March]. Available at: http://www.zmag.org/content/showarticle.cfm?ItemID=9951 [accessed: 8 August 2009].

Bigwood, J. 2008. New discoveries reveal US intervention in Bolivia. *Upside Down World* [Online, 14 October]. Available at: http://upsidedownworld.org/main/content/view/1522/1/ [accessed: 8 August 2009].

Blum, W. 2004. *Killing Hope: U.S. Military and CIA Interventions since World War II*. Monroe: Common Courage Press.

Bogdanich, W. and J. Nordberg. 2006. Mixed U.S. signals helped tilt Haiti toward chaos. *The New York Times*, 29 January.

Bolpress. 2008. El gobierno y organizaciones sociales del pacto de unidad en campaña por el Sí a la nueva constitución. *Rebelión* [Online, 3 May]. Available at: http://www.bolpress.com/art.php?Cod=2008030312 [accessed: 1 August 2009].

Bolton, M. *Human Security After Collapse: Global Security in Post-Earthquake Haiti*. LSE Global Governance Research Paper, RP 01, 2011.

Bonicelli, P. 2007. *USAIDs Strategy for Promoting Democracy and Building Markets in Latin America.* Lima: CIPE Conference.

Bronson, D. and L. Lamarche. 2001. *A Human Rights Framework for Trade in the Americas.* Rights and Democracy, March.

Brown, S. 2011. Aid effectiveness and the framing of new Canadian aid initiatives, in *Readings in Canadian Foreign Policy: Classic Debates and New Ideas*, edited by D. Bratt and C. Kukucha. Toronto: Oxford University Press.

Burdette, M. 1994. Structural adjustment and Canadian aid policy, in *Canadian International Development Assistance Policies: An Appraisal*, edited by C. Pratt. Kingston: McGill-Queen's University Press.

Burnell, P. 2008. From evaluating democracy assistance to appraising democracy promotion. *Political Studies*, 56 (2), 414–434.

Calloni, S. 2009. *Evo en la Mira: CIA y DEA en Bolivia.* Buenos Aires: Punto de Encuentro. Capítulo Boliviano de Derechos Humanos, Democracia y Desarrollo

Cameron, M. 2008. *Peru's Left and APRA's Victory* [Online]. Available at: http://www.politics.ubc.ca/fileadmin/user_upload/poli_sci/Graduate/Cdn_Com/Cameron_08.pdf [accessed: 28 July 2009].

Cameron, M. and E. Hershberg (editors). 2010. *Latin America's Left Turns: Politics, Policies, and Trajectories of Change.* Boulder: Lynne Rienner.

Canadian Council on American-Islamic Relations. 2010. *Write to Protest Nomination of Gerard Latulippe for President of NGO Rights and Democracy* [Online, 24 February]. Available at: http://www.caircan.ca/aa_more.php?id=3069_0_3_0_C [accessed: 1 March 2010].

Canadian Press. 2011. Prime Minister Stephen Harper in Latin America for six days to talk trade. *Huffington Post* [Online, July 8]. Available at: http://www.huffingtonpost.ca/2011/08/07/harper-latin-america-trade_n_920365.html [accessed: 9 July 2011].

Capítulo Boliviano de Derechos Humanos, Democracia y Desarrollo. 2009. *Pronunciamiento de las Organizaciones Defensoras de Derechos Humanos ante el Atentado contra la Vida del Director de CEJIS Responsable Regional del Beni* [Online]. Available at: http://www.cedib.org/bp/B20/documento9.pdf [accessed: 8 December 2009].

Carlsen, L. 2009a. *NAFTA's Dangerous Security Agenda* [Online: Americas Program, Center for International Policy, January 23]. Available at: http://www.cipamericas.org/archives/1583 [accessed: 4 February 2010].

Carlsen, L. 2009b. The sham elections in Honduras. *The Nation* [Online, 21 December]. Available at: http://www.thenation.com/article/sham-elections-honduras [accessed: 17 January 2010].

Carlsen, L. 2009c. *Victory in the Amazon* [Online: Americas Program, Center for International Policy, June 22]. Available at: http://www.cipamericas.org/archives/1745 [accessed: June 23, 2009].

Carlsen, L. 2011. *The Audacity of Free Trade Agreements* [Online: Americas Program, Center for International Policy]. Available at: http://www. cipamericas.org/archives/5102 [accessed: 8 December 2011].

Carlsen, L. 2012. *Doing Biden's Bidding* [Online: Americas Program, Center for International Policy]. Available at: http://www.cipamericas.org/archives/6496 [accessed: 5 March 2012].

Carothers, T. 2004. *Critical Mission: Essays on Democracy Promotion.* Washington DC: Carnegie Endowment for International Peace.

Carothers, T. 2006. The backlash against democracy promotion. *Foreign Policy*, March-April.

Carothers, T. 2009. *Democracy Promotion under Obama: Finding a way Forward* [Online: Carnegie Endowment for International Peace]. Available at: http://carnegieendowment.org/files/democracy_promotion_obama.pdf [accessed: 14 December 2011].

Carranza, M. 2009. The North–South Divide and security in the western hemisphere: United States–South American relations after September 11 and the Iraq war. *International Politics*, 46 (2), 276–297.

Carranza, M. 2010. Reality check: America's continuing pursuit of regional hegemony. *Contemporary Security Policy*, 31 (3), 406–440.

Carrarini, G. 2005. *Iniciativas de Alfabetización Intercultural Bilingüe: EL Rol de la Sociedad Civil en Bolivia.* Programa de Formación en Educación Intercultural Bilingüe para los Países Andinos. August.

Carroll, W. 2003. Undoing the End of History: Canada-Centred Reflections on the Challenge of Globalization, in *Global Shaping and its Alternatives*, edited by W. Carroll and Y. Atasoy. Bloomfield: Garamond Press.

Castonguay, A. Quand le politique s'arroge tous les droits, *Le Devoir*, 27–28 February.

Castor, S. and L. Garafola. 1974. The American occupation of Haiti (1915–34) and the Dominican Republic (1916–24). *The Massachusetts Review*, 15 (1), 253–275.

CBC News. 2004. Powell, Graham urge Aristide to end violence. *CBC News* [Online, 13 February]. Available at: http://www.cbc.ca/news/world/story/2004/02/13/haiti_police040213.html [accessed: 15 January 2006].

Center for Economic and Policy Research. 2011. *Haitian Companies Still Sidelined from Reconstruction Contracts* [Online, 19 April]. Available at: http://www. cepr.net/index.php/blogs/relief-and-reconstruction-watch/haitian-companies-still-sidelined-from-reconstruction-contracts [accessed: 24 January 2012].

Center for International Development. 2003. *Quarterly Report: Developing Skills of the Peruvian Congress* [Online: State University of New York, USAID report, 30 April]. Available at: http://pdf.usaid.gov/pdf_docs/PDABZ220.pdf [accessed: 17 June 2009].

Central Intelligence Agency. 2012. *The World Factbook* [Online]. Available at: https://www.cia.gov/library/publications/the-world-factbook/ [accessed: 23 January 2012].

Charbonneau, B. and W. Cox. 2008. Global order, US hegemony and military integration: the Canadian–American defense relationship. *International Political Sociology*, 2, 305–321.

Cheadle, B. 2010. Critics blast 'dead-of-night' Rights appointment. *The Globe and Mail*, March 3.

Chemonics International. 2003. *Bolivia Democratic Development and Citizen Participation: Final Report 1996–2003* [Online, November]. Available at: http://pdf.usaid.gov/pdf_docs/PDABZ607.pdf [accessed: 30 June 2009].

Chinai, R. 2008. *Haiti's Compounding Food and Health Crisis* [Online: Americas Program, Center for International Policy, 6 August]. Available at: http://americas.irc-online.org/am/5446 [accessed: 9 August 2008].

Chirapaq. 2006. *Las Mujeres Indígenas de Perú Exigen a los Candidatos Inclusión en sus Planes de Gobierno* [Online, 2 June]. Available at: http://www.aulaintercultural.org/article.php3?id_article=1628 [accessed: 8 July 2009].

CIDA. 1996. *Policy for CIDA on Human Rights, Democratization and Good Governance*. Gatineau, Quebec.

CIDA. 2006. *Taking a Second Look: The Potential of Peru's Oil and Gas Reserves* [Online, 1 December]. No longer available [accessed: 21 December 2008].

CIDA. 2007. *Programming Framework for Bolivia (2003–2007)* [Online]. No longer available [accessed: 20 December 2008].

CIDA. 2008a. *An Internal Guide for Effective Development Cooperation in Fragile States* [Online]. Gatineau.

CIDA. 2008b. *Bolivia Country Program Evaluation: Executive Report 2006* [Online, November]. Available at: http://www.acdi-cida.gc.ca/acdi-cida/acdi-cida.nsf/eng/NAT-22674521-GYF [accessed: 19 December 2008].

CIDA. 2009. *Disclosure of Grant and Contribution Awards Over $25,000* [Online]. Available at: http://www.acdi-cida.gc.ca/acdi-cida/contributions.nsf/rprts-eng?readForm [accessed: June–August, 2009].

CIDA. 2010a. *Canada Pledges Further Support for Relief, Recovery and Reconstruction in Haiti* [Online, 31 March]. Available at: http://www.acdi-cida.gc.ca/acdi-cida/ACDI-CIDA.nsf/eng/NAD-33195022-J92 [accessed: 24 June 2010].

CIDA. 2010b. *Haiti: CIDA Report* [Online]. Available at: http://www.acdi-cida.gc.ca/INET/IMAGES.NSF/vLUImages/Countries-of-Focus/$file/10-053-Haiti-E.pdf [accessed: 24 June 2010].

CIDA. 2010c. *Project Profile for Summit Follow-up by Civil Society – Phase II* [Online, January 4]. Available at: http://www.acdi-cida.gc.ca/cidaweb/cpo.nsf/fWebCAZEn?ReadForm [accessed: 8 June 2010].

CIDA. 2011. *Minister Oda Announces Initiatives to Increase the Benefits of Natural Resource Management for People in Africa and South America* [Online, 29 September]. Available at: http://www.acdi-cida.gc.ca/acdi-cida/ACDI-CIDA.nsf/eng/CAR-929105317-KGD [accessed: 2 March 2012].

CIDA. 2011. *Summary of Canada's Financial Contributions to Haiti in Response to the Earthquake* [Online, 11 January]. Available at: http://www.acdi-cida.gc.ca/acdi-cida/ACDI-CIDA.nsf/eng/FRA-4810272-JXY [accessed: 8 July].

Clement, C. 1997. Returning Aristide: The contradictions of US foreign policy in Haiti. *Race and Class*, 39 (2), 21–36.

Clement, C. 2005. Confronting Hugo Chavez: United States 'democracy promotion' in Latin America. *Latin American Perspectives*, 32 (3), May, 60–78.

Clinton, H. 2010. *Development in the 21st Century* [Online: Center for Global Development]. Available at: http://www.maximsnews.com/biohillaryclinton. htm [accessed: 2 December 2011].

Cohen, A. 2003. *While Canada Slept: How We Lost Our Place in the World.* Toronto: McClelland and Stewart.

Cole, N. 2007. Hugo Chavez and President Bush's credibility gap: The struggle against US democracy promotion. *International Political Science Review/ Revue internationale de science politique*, 28 (4), 493–507.

Collier, P. 2009. *Haiti: From Natural Catastrophe to Economic Security. A Report for the Secretary-General of the United Nations.* Department of Economics, Oxford University.

Collier, R. and J. Mahoney. 1997. Adding collective actors to collective outcomes: labor and recent democratization in South America and Southern Europe. *Comparative Politics*, 29 (3), 285–303.

Collins, M. 2010. Cuba: Democracy promotion programs under fire as fallout from spy arrest continues. *Upside Down World* [Online, 12 May]. Available at: http://upsidedownworld.org/main/cuba-archives-43/2488-cuba-democracy-promotion-programs-under-fire-as-fallout-from-spy-arrest-continues [accessed: 12 February 2012].

Concannon, Brian. 2006. *Haiti's Election: Right Result, for the Wrong Reasons* [Online: Jurist, University of Pittsburg, 17 February]. Available at: http:// jurist.law.pitt.edu/forumy/2006/02/haitis-election-right-result-for-wrong.php [accessed: 1 March 2006].

Consejo Permanente, OAS. 2009. *Informe de la Misión de Observación Electoral de la OEA sobre las Elecciones Generales y las Elecciones Presidenciales (2nda vuelta) Celebradas en la República del Perú el 9 de Abril y el 4 de Junio de 2006* [Online: CP/doc. 4428/09, 25 August]. Available at: http://www.oas. org/sap/docs/misiones/2006/CP%20Informe%20MOE%20Per%C3%BA%20 2006.pdf [accessed: 18 July 2009].

CONSODE. 2003. *Ojo Ciudadano en el Congreso: Boletín informativo del CONSODE* [Online: CONSODE]. Lima: CONSODE.

CONSODE. 2004. *Apoyo a la Sociedad Civil para Reformas Politicas Democraticas en el Peru: Congreso.* Lima: CONSODE.

Cooper, A. and T. Legler. 2005. A tale of two mesas: The OAS defense of democracy in Peru and Venezuela. *Global Governance*, 11, 425–444.

Cooper, A., R. Higgott, and K. Nossal. 1993. *Relocating Middle Powers: Australia and Canada in a Changing World Order.* Vancouver: UBC Press.

Coughlin, D. and K. Ives. 2011a. WikiLeaks Haiti: Let them live on $3 a day. *The Nation* [Online, 1 June]. Available at: http://www.thenation.com/ article/161057/wikileaks-haiti-let-them-live-3-day [accessed: 3 June 2011].

Coughlin, D. and K. Ives. 2011b. WikiLeaks Haiti: The PetroCaribe files. *The Nation* [Online, 1 June]. Available at: http://www.thenation.com/ article/161056/petrocaribe-files [accessed: 3 June 2011].

The Council of Canadians. 2009. *Canadians Ask Federal Government to Halt Ratification of Canada-Peru Free Trade Agreement in Light of Peruvian Police Massacre of Indigenous Protestors* [Online: Press release]. Available at: http://www.canadians.org/media/trade/2009/11-June-09.html [accessed: 14 October 2011].

Cox, M. 2000. Wilsonianism resurgent? The Clinton administration and American democracy promotion in the late 20th century, in *American Democracy Promotion: Impulses, Strategies, Impacts*, edited by M. Cox, G.J. Ikenberry, and T. Inoguchi. Oxford: Oxford University Press.

Cox, R. 1993. Gramsci, hegemony and international relations: An essay in method, in *Gramsci, Historical Materialism and International Relations*, edited by S. Gill. Cambridge: Cambridge University Press.

Cox, R. 2001. Civil Society at the turn of the millennium: Prospects for an alternative world order. *Review of International Studies*, 25 (1), 3–28.

Cox, R. 2005. A Canadian dilemma: The United States or the world. *International Journal*, 60 (3), Summer.

Cox, R. and M. Schechter. 2002. *The Political Economy of a Plural World: Critical Reflections on Power, Morals and Civilization.* London: Routledge.

Cox, R. and T. Sinclair. 1996. *Approaches to World Order.* Cambridge: Cambridge University Press.

CPESC. 2008. *Public Denouncement: Assault, Looting and Destruction of the Headquarters of the Indigenous People of Santa Cruz-Bolivia* [Online: Ukhampacha Bolivia, Santa Cruz de la Sierra, Bolivia]. No longer available [accessed: 8 July 2009].

Craig, D. and D. Porter. 2006. *Development Beyond Neoliberalism?: Governance, Poverty Reduction and Political Economy.* London: Routledge.

Craner, L. and K. Wollack. 2008. *Don't go it Alone: America's Interest in International Cooperation. New Directions for Democracy Promotion* [Online: IRI and NDI; Better World Campaign, 24 July]. Available at: http:// www.ndi.org/files/2344_newdirections_engpdf_07242008.pdf [accessed: 30 June 2009].

Dade, C. 2008. The Monroe Doctrine is dead: Long live Canada. *Miami Herald*, 2 August.

Dagnino, E., A. Olvera and A. Panfichi. 2008. Democratic Innovation in Latin America: A First Look at the Democratic Participatory Project, in *Democratic Innovations in the South*, edited by Dagnino, E., A. Olvera and A. Panfichi. Buenos Aires: Consejo Latinoamericano de Ciencias Sociales.

Dahl, R. 1971. *Polyarchy, Participation and Opposition.* New Haven: Yale University Press.

Dahl, R. 1985. *A Preface to Economic Democracy.* Cambridge: Polity Press.

Dangl, B. 2007. *The Price of Fire: Resource Wars and Social Movements in Bolivia.* Edinburgh: AK Press.

Dangl, B. 2008. Undermining Bolivia: A landscape of Washington intervention. *The Progressive* [Online, February]. Available at: http://upsidedownworld. org/main/bolivia-archives-31/1124-undermining-bolivia-a-landscape-of-washington-intervention [accessed: 14 April 2009].

Deere, C. and M. León. 2001. Who owns the land? Gender and land-titling programmes in Latin America. *Journal of Agrarian Change*, 1 (3), 440–467.

Defensoría del Pueblo. 2007. *Executive Summary of the Tenth Annual Report of the Ombudsman's Office, January – December 2006.* Lima: República del Perú: Defensoría del Pueblo, Ombudsman's Office Peru.

Deibert, M. 2005. *Notes from the Last Testament: The Struggle for Haiti.* New York: Seven Stories Press.

Department of Commerce. 2009. *The United States Contributes to Economic Prosperity in Peru* [Online]. Available at: http://www.trade.gov/ promotingtrade/westhemprosperity/peru.pdf [accessed: 20 November 2009].

Development and Peace. 2003. *Support for the Processes of Democratization: 2000– 2003 Integrated Program, Triennial Report on Results* [Online]. Available at: http://www.devp2.org/devpme/documents/eng/pdf/ReportResults-Nov2003. pdf [accessed: 1 May 2009].

Development and Peace. 2006a. *Support for the Democratization of Development: 2003–2006* [Online]. Available at: http://www.devp2.org/devpme/eng/ publications/program0306-eng.html [accessed: 1 May 2009].

Development and Peace. 2006b. *Integrated Program 2006–2011: Building More Just and Equitable Communities* [Online]. Available at: http://www.devp2. org/devpme/eng/publications/documents/2006-2011PROGRAM_ENG.pdf [accessed: 1 May 2009].

Diamond, L. 1999. *Developing Democracy: Toward Consolidation.* Baltimore: The Johns Hopkins University Press.

Diamond, L. 2011. *Democracy Promotion and the Obama Doctrine* [Online: Council on Foreign Relations]. Available at: http://www.cfr.org/us-strategy-and-politics/democracy-promotion-obama-doctrine/p24621 [accessed: 1 November 2011].

DFAIT. 2007. *A New Focus on Democracy Support: Government Response to the Eight Report of the Standing Committee on Foreign Affairs and International Development* [Online, 2 November]. Available at: http://www.parl.gc.ca/ HousePublications/Publication.aspx?DocId=3093769&Language=E&Mode= 1&Parl=39&Ses=1 [accessed: 20 January 2008].

DFAIT. 2008. *Glyn Berry Program by Thematic Focus* [Online, 5 November]. Available at: http://www.international.gc.ca/glynberry/program-gb-programme/thematic-theme.aspx?lang=eng#democratic [accessed: 8 December 2011].

DFAIT. 2009. *Building the Canadian Advantage: A Corporate Social Responsibility (CSR) Strategy for the Canadian International Extractive*

Sector [Online, March]. Available at: http://www.international.gc.ca/trade-agreements-accords-commerciaux/ds/csr-strategy-rse-stategie.aspx [accessed: 31 October 2009].

DFAIT. 2011a. *Address by Minister of State Ablonczy at Inter-American Democratic Charter Commemoration Event* [Online: Valparaiso, Chile, 5 September]. Available at: http://www.international.gc.ca/media/state-etat/speeches-discours/2011/028.aspx?lang=eng&view=d [accessed: 4 March 2012].

DFAIT. 2011b. *Seizing Global Advantage* [Online, 8 August]. Available at: http://www.international.gc.ca/commerce/strategy-strategie/r12.aspx?view=d [accessed: 1 December 2011].

DFAIT. 2011c. *Canada Contributes to Improving Security in Americas* [Online, 7 June]. Available at: http://www.international.gc.ca/media/aff/news-communiques/2011/154.aspx?lang=eng&view=d [accessed: 13 March 2012].

DFAIT. 2011d. Evaluation of the Americas Strategy [Online]. Available at: http://www.international.gc.ca/about-a_propos/oig-big/2011/evaluation/tas_lsa11.aspx?lang=engandview=d [accessed: 12 November 2011].

Dhaliwal, A. 1996. Can the subaltern vote? Radical democracy, discourses of representation and rights, and questions of race, in *Radical Democracy: Identity, Citizenship, and the State*, edited by D. Trend. New York: Routledge, 42–61.

Dobriansky, P. 2006. *Democracy in Latin America: Successes, Challenges and the Future. Statement by Paula J. Dobriansky, Under Secretary for Democracy and Global Affairs, before the House International Relations Committee* [Online]. Available at: http://commdocs.house.gov/committees/intlrel/hfa28366.000/hfa28366_0f.htm [accessed: 18 June 2009].

Drzewieniecki, J. 2002. La Coordinadora Nacional De Derechos Humanos De Perú: un estudio de caso, in *Sociedad Civil, Esfera Pública y Democratización en América Latina: Andes y Cono Sur* edited by Aldo Panfichi. Lima: Universidad Católica del Perú, Fondo de Cultura Económica, 516–547.

Dupuy, A. 1997. *Haiti in the New World Order: The Limits of the Democratic Revolution.* Boulder: Westview Press.

Dupuy, A. 2005. From Jean-Bertrand Aristide to Gerard Latortue: The unending crisis of democratization in Haiti. *Journal of Latin American Anthropology*, 10 (1), 186–205.

Dupuy, A. 2007. *The Prophet and Power: Jean-Bertrand Aristide, the International Community, and Haiti.* Lanham: Rowman and Littlefield Publishers.

Dupuy, A. 2010. Commentary Beyond the Earthquake: A Wakeup Call for Haiti. *Latin American Perspectives*, 37(3), 195–204.

Eaton, K. 2007. Backlash in Bolivia: Regional autonomy as a reaction against indigenous mobilization. *Politics and Society*, 35 (1), 71–102.

Echave, J. 2005. *Canadian Mining Companies Investments in Peru: The Tambogrande Case and the Need to Implement Reforms* [Online: CooperAccion]. No longer available [accessed: September 4, 2009].

ECLAC. 2003. *Canada's Trade and Investment with Latin America and the Caribbean.* ECLAC: Santiago.

ECLAC. 2004. *Social Panorama of Latin America*: 2002–2003. ECLAC: Santiago.

ECLAC. 2009. *Quality of Latin American and Caribbean Industrialization and Integration into the Global Economy*. ECLAC: Santiago.

ECLAC. 2010a. *Time for Equality: Closing Gaps, Opening Trails*. ECLAC: Santiago.

ECLAC. 2010b. *Foreign Direct Investment in Latin America and the Caribbean*. ECLAC: Santiago.

Economist Intelligence Unit. 2007. *Country Report: Peru.* September.

Economist Intelligence Unit. 2009. *Country Report: Peru.* September.

Edmonds, K. 2012. Canada: A late but eager partner in policing the Caribbean. *NACLA Report on the Americas* [Online, 2 February]. Available at: http://nacla.org/blog/2012/2/2/canada-late-eager-partner-policing-caribbean-1 [accessed: 3 February 2012].

Emersberger, J. 2007. January 19. *Amnesty International's Track Record in Haiti since 2004* [Online: Haitianalysis.com]. Available at: http://www.haitianalysis.com/2007/1/19/amnesty-international%E2%80%99s-track-record-in-haiti-since-2004 [accessed: 8 February 2007].

Emerson, G. 2010. Radical neglect? The 'war on terror' and Latin America. *Latin American Politics and Society*, 52(1), 33–52.

Encarnación, O. 2002. Venezuela's civil society coup. *World Policy Journal*, Summer, 38–48.

Engler, Y. 2008. *Quebec and Haiti* [Online: ZSpace, April 14]. Available at: http://www.zmag.org/znet/viewArticle/17145 [accessed: 14 June 2009].

Engler, Y. 2009. *The Black Book of Canadian Foreign Policy.* Black Point: Fernwood Publishing.

Engler, Y. and A. Fenton. 2005. *Canada in Haiti: Waging War on the Poor Majority.* Vancouver: Red Publishing.

Epstein, S. and K. Nakamura. 2009. *State, Foreign Operations and Related Programs: FY 2009 Appropriations* [Online: CRS Report for Congress, 7-5700, 3 April]. Available at: http://www.fas.org/sgp/crs/row/RL34552.pdf [accessed: 28 August 2009].

Erikson, D. The Obama administration and Latin America: Towards a new partnership?, in *Inter-American Cooperation at a Crossroads*, edited by G. Mace, A. Cooper and T. Shaw. London: Palgrave Macmillan, 43 59.

Estevadeordal, A. and K. Suominen. Economic integration in the Americas: An unfinished agenda, in *Inter-American Cooperation at a Crossroads*, edited by G. Mace, A. Cooper and T. Shaw. London: Palgrave Macmillan, 81–94

Ethier, D. 2003. Is democracy promotion effective? Comparing conditionality and incentives. *Democratization*, 10 (1), 99–120.

Faqundez, F. 2009. Morales denuncia que la USAID financia campaña de opositores. *YVKE Mundial: Internacionales* [Online, 6 September 6]. No longer available [accessed: 24 October 2009].

Farmer, P. 2004. Who Removed Aristide? *London Review of Books*, 26 (8), 28–31 April.

Farmer, P. 2011. *Haiti after the Earthquake*. New York: Public Affairs.

Fatton, R. 2002. *Haiti's Predatory Republic: The Unending Transition to Democracy*. Boulder: Lynne Rienner Publishers.

Fatton, R. 2006. The fall of Aristide and Haiti's current predicament, in *Haiti: Hope for a Fragile State*, edited by Y. Shamsie and A. Thompson. Waterloo: Wilfrid Laurier University Press, 15–24.

Fauriol, G. 2006. *Remarks by Georges Fauriol to the World Affairs Council: The Role of Elections in Expanding Freedom and Building Democracy Worldwide* [Online: Minneapolis]. Available at: http://www.iri.org/sites/default/files/2006-04-18-FauriolRemarks.pdf [accessed: 14 June 2009].

FBDM. 2007. *Los Consensos de la Constituyente*. La Paz: Temas de Réflexion, 17. November.

Fenton, A. 2005. *The Canadian Corporate/ State Nexus in Haiti* [Online: ZSpace, 16 May]. Available at: http://www.zmag.org/znet/viewArticle/6264 [accessed: 4 March 2006].

Flynn, L., R. Roth and L. Fleming. 2004. *Report of the Haiti Accompaniment Project* [Online, 29 June–9 July]. Available at: http://www.haitiaction.net/News/hap6_29_4.html [accessed: 8 September 2006].

Flynn, M. 2007. Between subimperialism and globalization: a case study in the internationalization of Brazilian capital. *Latin American Perspectives*, 34 (6), 9–27.

FM Bolivia. 2009. *El Movimiento Bolivia Libre (MBL) se Adhiere al Movimiento Al Socialismo (MAS) y Respalda el Proceso de Cambio* [Online, 4 November]. Available at: http://www.fmbolivia.com/noticia18919-elmovimiento-bolivia-libre-mbl-se-adhiere-al-socialismo-mas-y-respada-elproceso-de-cambio.html [accessed: 1 December 2009].

Forero, J. 2006. Seeking united Latin America, Venezuela's Chávez is a divider. *New York Times* [Online, 19 May]. Available at: http://www.nytimes.com/2006/05/20/world/americas/20chavez.html?_r=1&scp=9&sq=alejandro+toledo+bush&st=nyt [accessed: 13 March 2012].

Fox, M. 2009. ALBA summit ratifies regional currency, prepares for Trinidad. *Upside Down World* [Online, 20 April]. Available at: http://upsidedownworld.org/main/content/view/1819/68/ [accessed: 4 June 2009].

Frank, D. 2011. Open season on teachers in Honduras. *The Nation* [Online, 5 May]. Available at: http://www.thenation.com/article/160472/open-season-teachers-honduras?comment_sort=ASC [accessed: 3 December 2011].

Freedom House, Radio Free Europe/ Radio Liberty and Radio Free Asia. 2009. *Undermining Democracy: 21st Century Authoritarians*. Washington DC: June.

French, J. 2010. Many lefts, one path? Chávez and Lula, in *Latin America's Left Turns: Politics, Policies, and Trajectories of Change*, edited by M. Cameron and E. Hershberg. Boulder: Lynne Rienner, 41–60.

Fuentes, F. 2011. Government, social movements, and revolution in Bolivia today: A response to Jeffery Webber. *International Socialist Review* [Online, 76, March–April]. Available at: http://isreview.org/issues/76/debate-bolivia.shtml [accessed: 23 May 2011].

Galeano, E. 1973. *Open Veins of Latin America: Five Centuries of the Pillage of a Continent*. New York: Monthly Review Press.

Gallagher, K. 2010. What's left for Latin America to do with China? *NACLA Report on the Americas*, 43 (3), 50–53.

Gamarra, E. 2003. *Conflict Vulnerability Assessment Bolivia* [Online: Latin America and Caribbean Center, Florida International University. Report Commissioned by USAID]. Available at: http://pdf.usaid.gov/pdf_docs/PNADF145.pdf [accessed: 20 June 2009].

Gamble, A. and A. Payne. 1996. *Regionalism and World Order*. New York: St. Martin's Press.

Gershman, C. 2000. *The Role of Civil Society Organizations in the Global Movement for Democracy* [Online: The Conference on Challenges for Civil Society in the Emerging World Order, Valencia, Spain, 27 November]. Available at: http://www.ned.org/about/board/meet-our-president/archived-remarks-and-presentations/112700 [accessed: 18 July 2009].

Gershman, C. 2008. *Remarks to the First Meeting of the Assembly of the Latin American Network for Democracy* [Online: Panama City: 29 February]. Available at: http://www.ned.org/about/board/meet-our-president/archived-remarks-and-presentations/022908 [accessed: July 18, 2009].

Gershman, C. and M. Allen. 2006. The Assault on Democracy Assistance. *Journal of Democracy*, 17 (2), 36–51.

Goff, S. and A. Fenton. 2004. The invasion of Haiti: Interview with Stan Goff. *Z-Net* [Online, 19 May]. Available at: http://www.zmag.org/znet/viewArticle/8523 [accessed: 5 March 2006].

Golinger, E. 2009. Washington and the coup in Honduras: Here is the evidence. *mediaLeft* [Online, 15 July]. Available at: http://medialeft.net/main/index.php/voices-medialeftsections-142/80-eva-golinger/1212-washington-a-the-coup-in-honduras-here-is-the-evidence [accessed: 4 December 2009].

Gordon, T. 2010. *Imperialist Canada*. Winnipeg: Arbeiter Ring.

Gordon, T. and J. Webber. 2008. Imperialism and resistance: Canadian mining companies in Latin America. *Third World Quarterly*, 29 (1), 63–87.

Gordon, T. and J. Webber. 2010. Canada's Long Embrace of the Honduran Dictatorship. *Counterpunch* [Online, 19–21 March]. Available at: http://www.counterpunch.org/2010/03/19/canada-s-long-embrace-of-the-honduran-dictatorship/ [accessed: 22 March 2010].

Gordon, T. and J. Webber. 2011. Canada and the Honduran coup. *Bulletin of Latin American Research*, January 10, 30, 328–342.

Government of Canada. 2007. *Government Response to the Eight Report of the Standing Committee on Foreign Affairs and International Development: A New Focus on Democracy Support* [Online]. Available at: http://www2.parl.gc.ca/HousePublications/Publication.aspx?DocId=3093769andLanguage=EandMode=1andParl=39andSes=1 [accessed: 18 June 2009].

Government of Canada. 2008. *Canada – Bolivia Relations* [Online, 11 December]. Available at: http://www.canadainternational.gc.ca/peru-perou/bilateral_

relations_bilaterales/canada_bolivia-bolivie.aspx?menu_id=8andmenu=L [accessed: 20 December 2008].

Government of Canada. 2009a. *Fact Sheet: Haiti* [Online, 16 July]. Available at: www.haiti.gc.ca [accessed: 20 July 2011].

Government of Canada. 2009b. *Canada – Peru Relations* [Online, 19 June]. Available at: http://www.canadainternational.gc.ca/peru-perou/bilateral_relations_bilaterales/canda_peru-perou.aspx?menu_id=7andmenu=L [accessed: 18 December 2011].

Government of Canada. 2009c. *Canada Promotes Socially Responsible Mining in Peru* [Online, 20 May]. Available at: http://www.canadainternational.gc.ca/peru-perou/highlights-faits/PERCAN2009.aspx?lang=eng [accessed: 18 December 2011].

Gramsci, A., Q. Hoare and G. Nowell-Smith. 1971. *Selections from the Prison Notebooks of Antonio Gramsci*. New York: International Publishers.

Grandin, G. 2007. Democracy, Diplomacy and Intervention in the Americas. *NACLA Report on the Americas*, 40 (1), January–February, 22–25.

Grant, T. 2011. Canadian companies in Haiti, before and after the quake. *Globe and Mail* [Online, 11 January]. Available at: http://www.theglobeandmail.com/news/world/project-jacmel/haiti-one-year-later/canadian- December, 2011].

Griffin, T. 2004. *Haiti, Human Rights Investigation: November 11–21, 2004* [Online]. Available at: http://ijdh.org/CSHRhaitireport.pdf [accessed: 14 November 2005].

Gros, J.G. 2010. Anatomy of a Haitian tragedy: When the fury of nature meets the debility of the state. *Journal of Black Studies*, 42 (2), 131–157.

The Guardian. 2011. US embassy cables: Mining companies worried about security. *The Guardian* [Online, 31 January]. Available at: http://www.guardian.co.uk/world/us-embassy-cables-documents/38881 [accessed: 17 February].

Guilhot, N. 2005. *The Democracy Makers: Human Rights and the Politics of Global Order*. New York: Columbia University Press.

Gurzu, A. 2010. Haitian Elections a Frustration for Donors: Cable Canada among Countries Worried about Fairness. *Embassy*, June 29.

Haiti Democracy Project. 2007. Mission Statement [Online]. Available at: http://www.haitipolicy.org/main/mission_statement.htm [accessed: 18 June 2009].

Hallward, P. 2004. Option Zero in Haiti. *New Left Review*, 27, May–June, 23–48.

Hallward, P. 2007. *Damming the Flood: Haiti, Aristide, and the Politics of Containment*. London: Verso.

Hallward, P. 2008. Haiti debate: Peter Hallward responds to Michael Deibert's review of Damming the Flood. *Monthly Review* [Online, 18 July]. Available at: http://mrzine.monthlyreview.org/ [accessed: 6 July 2009].

Harris, J. 2007. Bolivia and Venezuela: The democratic dialectic in new revolutionary movements. *Race and Class*, 49 (1), 1–24.

Herz, A. and K. Ives. 2011. WikiLeaks Haiti: The post-quake 'gold rush' for reconstruction contracts. *The Nation* [Online, 15 June]. Available at: http://

www.thenation.com/article/161469/wikileaks-haiti-post-quake-gold-rush-reconstruction-contracts [accessed: 14 June 2011].

Hill, G., K. McBride and J. Diaz-Albertini. 2003. *Final Evaluation of OTI's Program in Peru* [Online: USAID, Bureau for Democracy, Conflict and Humanitarian Assistance, Office of Transition Initiatives]. Available at: http://www.usaid.gov/our_work/cross-cutting_programs/transition_initiatives/country/peru/rptFinal.pdf [accessed: 3 December 2008].

Hordijk, M. 2005. Participatory governance in Peru: Exercising citizenship. *Environment and Urbanization*, 17 (1), 219–236.

Hornbeck, J.F. 2011. U.S.–Latin American Trade: Recent Trends and Policy Issues [Online: Congressional Research Service, Report for Congress, 8 February]. Available at: http://www.fas.org/sgp/crs/row/98-840.pdf [accessed: 18 December 2011].

House of Commons, Canada. 2004. *Hansard*. Ottawa: 37th Parliament, 3rd session (023), March 10.

House of Commons, Canada. 2007. *Advancing Canada's Role in International Support for Democratic Development.* Standing Committee on Foreign Affairs and International Development. Ottawa: 39th Parliament, 1st Session, July.

House of Representatives. 2004. *United States Support of Human Rights and Democracy.* One Hundred and Eight Congress, Second Session. Washington DC.

Hughes, N. 2010. Indigenous protest in Peru: The "orchard dog" bites back.' *Social Movements Studies*, 9 (1), 85–90.

Hylton, F. 2008. Reactionary rampage: The paramilitary massacre in Bolivia. *NACLA Report on the Americas* [Online, 16 September]. Available at: http://nacla.org/node/5021 [accessed: 18 September 2008].

IFES. 2004. *Final Report: Executive Summary, Haiti Constituency Building for Judicial Reform* [Online]. Available at: http://www.ifes.org/pubsearch_results.html?region_name_0=Haitiandandandandkeyword=andandt=1 [accessed: 8 March 2006].

Ignatieff, M. 2003. *Empire Lite: Nation-Building in Bosnia, Kosovo and Afghanistan.* Toronto: Penguin Canada.

Ingalls, L. 2006. *Haitians Brave Large Crowds, Delays to Vote* [Online: IFES, 8 February]. Available at: http://www.ifes.org/features.html?title=Haitians%20Brave%20Large%20Crowds,%20Delays%20to%20Vote [accessed: 8 March 2006].

Institute for Justice and Democracy in Haiti. 2010. The international Community Should Pressure the Haitian Government for Prompt and Fair Elections. [Online, 30 June]. Available at: http://ijdh.org/archives/13138 [accessed: 6 June 2011].

Institute for Justice and Democracy in Haiti. 2010. *Haiti's November 28 Elections: Trying to Legitimize the Illegitimate* [Online, 22 November]. Available at: http://ijdh.org/archives/15456 [accessed: 6 June 2011].

Institute for Justice and Democracy in Haiti and *Bureau des Avocats Internationaux.* 2011. *Football over Families: Human Rights Groups Denounce Haitian*

Government's Unlawful Eviction of Homeless Earthquake Survivors from Stadium [Online: Press Release, 21 July]. Available at: http://ijdh.org/archives/20260 [accessed: 6 June 2011].

Institute for Justice and Democracy in Haiti. 2004. *Human Rights Violations in Haiti* [Online, February–March]. Available at: http://ijdh.org/articles/article_ijdh-human-rights-violations.php [accessed: 11 March 2006].

International Commission on Intervention and State Sovereignty. 2001. *The Responsibility to Protect.* Ottawa: International Development Research Centre. December.

IRI. 2005a. *Haiti Political Party Development* [Online: USAID, quarterly report, April–June]. Available at: http://pdf.usaid.gov/pdf_docs/PDACF866.pdf [accessed: 1 April 2006].

IRI. 2005b. *Peru: Promoting Political Stability by Improving Government Communications* [Online: USAID, quarterly report, April–June]. Available at: http://mediamapresource.wikispaces.com/file/view/Promoting+Political+Stability+by+Improving+Government+Communications.pdf [accessed: 1 June 2009].

IRI. 2005c. *Bolivia: Improving Citizen Perceptions of Political Parties* [Online: USAID, final quarterly report, April–June]. Available at: http://pdf.usaid.gov/pdf_docs/PDACF865.pdf [accessed: 1 June 2009].

IRI. 2008a. *Advancing Democracy: A Report of The International Republican Institute, IRI Honours Salvadoran President* [Online: IRI Report, Winter – Spring]. Available at: http://www.iri.org/sites/default/files/2008%20Vol.%2019%20Issue%201.pdf [accessed: 1 June 2009].

IRI. 2008b. *Facts about IRI's Work in Haiti* [Online]. Available at: http://www.iri.org/newsreleases/2008-07-18-Haiti-faq.asp [accessed: 1 June 2009].

Ives, K. and A. Herz. 2011. WikiLeaks Haiti: The Aristide Files. *The Nation* [Online, 5 August]. Available at: http://www.thenation.com/article/162598/wikileaks-haiti-aristide-files?page=0,1 [accessed: 8 August 2011].

Ivison, J. 2009. Funding for leftist group to be cut. *The National Post* [Online, 5 December]. Available at: http://www.nationalpost.com/opinion/columnists/story.html?id=2e5f8e01-984e-4b95-91a3-f0a50a4afee8 [accessed: 5 December 2009].

Jacobsen, J. 2009. *Charges of U.S. Funding to Violent Opposition Groups in Bolivia: The National Endowment for Democracy Responds* [Online, The Democracy Center, 8 May]. Available at: http://democracyctr.org/blog/2009/05/charges-of-us-funding-to-violent.html [accessed: 22 June 2009].

Jean, J.C. and M. Maesschalck. 1999. *Transition Politique En Haïti: Radiographie Du Pouvoir Lavalas.* Paris and Montreal: L'Harmattan.

Jessop, B. and N.L. Sum. 2001. Pre-disciplinary and post-disciplinary perspectives. *New Political Economy*, 6 (1), 89–101.

Johnston, J. and M. Weisbrot. 2011. *Haiti's Fatally Flawed Election* [Online: Center for Economic and Policy Research, January]. Available at: http://

www.cepr.net/index.php/publications/reports/haitis-fatally-flawed-election [accessed: 1 February 2011].

Jorge Legoas, P. 2007. 'Watchdogs'. Ciudadanía y discursos del desarrollo. *Tabula Rasa*, (7), July–December, 17–46.

Joyce, R. 2010. Legitimizing the illegitimate: The Honduran show elections and the challenge ahead. *NACLA Report on the Americas*, 43 (2).

Jubilee USA Network. 2010. *IMF Takes Two Steps Forward and One Step Back on Haiti* [Online, Press Release, 22 July]. Available at: http://www.jubileedebtcampaign.org.uk/?lid=5460 [accessed: 1 August 2011].

Karatnycky, A. 2004. The democratic imperative. *The National Interest*, summer.

Keenan, K. 2010. Canadian mining: Still unaccountable. *NACLA Report on the Americas*, 43 (3), 29–34.

Klein, N. 2007. *The Shock Doctrine: The Rise of Disaster Capitalism*. New York: Metropolitan Books/Henry Holt.

Kohl, B. and L. Farthing. 2006. *Impasse in Bolivia: Neoliberal Hegemony and Popular Resistance*. London: Zed Books.

Kolbe, A. and R. Hutson. 2006. *Human Rights Abuses and other Criminal Violations in Port-au-Prince, Haiti: A Random Survey of Households*. The Lancet [Online, 31 August]. Available at: http://ijdh.org/pdf/Lancet%20 Article%208-06.pdf [accessed: 4 October 2006].

Kopstein, J. 2006. *EU, American, and Canadian Approaches to Democracy Promotion: Are they Compatible?* [Online: workshop, Austrian Embassy Ottawa and Foreign Affairs Canada]. Available at: http://canada-europe-dialogue.ca/events/Workshop-June12-2006/KopsteinEU-US-Canada-DemocracyPromotion.pdf [accessed: 30 November 2006]

Kristoff, M. and L Panarelli. 2010. *Haiti: A Republic of NGOs?* [Online: United States Institute of Peace Brief, 23]. Available at: http://www.usip.org/publications/haiti-republic-ngos [accessed: 14 July 2011].

Kuyek, J. 2006. Legitimating plunder: Canadian mining companies and corporate social responsibility, in *Community Rights and Corporate Responsibility: Canadian Mining and Oil companies in Latin America*, edited by L. North, V. Patroni and T. Clark. Toronto: Between the Lines, 203–221.

La Razón. 2006. *Bolivia Asamblea Constituyente y Autonomias: Pacto de Unidad Presiona* [Online]. No longer available [accessed: 8 August 2009]

Laplante, L. 2008. Transitional justice and peace building: Diagnosing and addressing the socioeconomic roots of violence through a human rights framework. *International Journal of Transitional Justice*, 2 (3), 331–355.

Legler, T. 2006. *Bridging Divides, Breaking Impasses: Civil Society and the Protection and Promotion of Democracy in the Americas* [Online: FOCAL, FP-06-02]. Available at: http://www.civil-society.oas.org/News/FOCAL%20 -%20Thomas%20Legler%20-%20English.pdf [accessed: 28 June 2009].

Legler, T. 2011. Demise of the inter-American democracy promotion regime?, in *Inter American Cooperation at a Crossroads*, edited by G. Mace, A. Cooper and T. Shaw. London: Palgrave Macmillan, 111–120.

Leogrande, W. 1998. *Our Own Backyard – The United States in Central America, 1977–1992*. Chapel Hill: University of North Carolina Press.

Levitsky, S. and M. Cameron. 2003. Democracy without parties? Political parties and regime change in Fujimori's Peru. *Latin American Politics and Society*, 10 (27), 1–33.

Leysens, A. 2008. *The Critical Theory of Robert W Cox: Fugitive or Guru?* Houndmills, Basingstoke: Palgrave Macmillan.

Lindsay, R. 2005. Exporting gas and importing democracy in Bolivia. *NACLA Report on the Americas*, 39 (3), 5–11.

Long, S. 2004. *Aiding Oppression in Haiti: Kofi Annan and General Heleno's Complicity in Latortue's Jackal Regime* [Online: Council on Hemispheric Affairs, 16 December]. Available at: http://www.coha.org/aiding-oppression-in-haiti-kofi-annan-and-general-heleno%E2%80%99s-complicity-in-latortue%E2%80%99s-jackal-regime/ [accessed: 18 March 2006].

Lopez Maya, M. 2007. Venezuela today: a 'participative and protagonistic' democracy?, in *Socialist Register 2008, Global Flashpoints: Reactions to Imperialism and Neoliberalism*, edited by L. Panitch and C. Leys. Monmouth: The Merlin Press, 160–180.

Lowe, D. 2008. *Idea to Reality: A Brief History of the National Endowment for Democracy* [Online: National Endowment for Democracy, 27 July]. Available at: http://www.ned.org/about/nedhistory.html [accessed: 30 July 2009].

Macdonald, L. 1994. Globalising civil society: Interpreting international NGOs in Central America. *Millennium*, 23 (2), 267–285.

Macdonald, L. 1995. Unequal partnerships: The politics of Canada's relations with the Third World. *Studies in Political Economy*, 47, Summer, 111–141.

Mace, G., A. Cooper and T. Shaw. 2011. Introduction, in *Inter-American Cooperation at a Crossroads*, edited by G. Mace, A. Cooper and T. Shaw. London: Palgrave Macmillan, 1–22.

Main, A. 2010. USAID: The Bone of Contention in U.S.–Bolivia Relations [Online: Center for Economic and Policy Research, 22 June]. Available at: http://www.cepr.net/index.php/blogs/cepr-blog/usaid-bone-ofcontention-in-us-bolivia-relations/ [accessed: 14 May 2011].

Maloney, S. 2000. Maple leaf over the Caribbean: Gunboat diplomacy Canadian style? In *Canadian Gunboat Diplomacy: The Canadian Navy & Foreign Policy*, edited by A. Griffiths, R. Gimblett, and P. Haydon. Halifax: Centre for Foreign Policy Studies.

Management Systems International. 2000. *Evaluation of USAID/ Peru's Democracy Education Activities: Final Report* [Online: USAID, April]. Available at: http://www.gci275.com/archive/demo_edu.pdf [accessed: 16 June 2009].

Martin, L. 2010. *Harperland: The Politics of Control*. Toronto: Viking Canada.

Matthews, R. and C. Pratt. 1988. *Human Rights in Canadian Foreign Policy*. Kingston: McGill-Queen's University Press.

McDermott, J. 2009. Toxic fallout of Colombian scandal. *BBC News* [Online, 7 May]. Available at: http://news.bbc.co.uk/2/hi/uk_news/8038399.stm [accessed: 1 November 2011].

McIntosh Sundstrom, L. 2005. Hard choices, good causes: Exploring options for Canada's overseas democracy assistance. *IRPP Policy Matters*, 6 (4), September.

McNeish, J.A. 2008. Constitutionalism in an insurgent state: Rethinking the legal empowerment of the poor in a divided Bolivia, in *Rights and Legal Empowerment in Eradicating Poverty*, edited by D. Banik. Farnham: Ashgate Publishing, 69–96.

McSherry, P. 2005. Operation Condor as a hemispheric 'counterterror' organization, in *When States Kill: Latin America, the U.S., and Technologies of Terror*, edited by N. Rodriguez and C. Menjívar. Austin: University of Texas Press, 28–58.

Meléndez López, L. and P. Sarmiento Rissi. 2008. *National Report on Feminicide in Peru* [Online: CLADEM]. Available at: http://www.demus.org.pe/publicacion/022_doc_report_fem_peru_ingles.pdf [accessed: 23 November 2009].

Melia, T. 2006. The democracy bureaucracy. *The American Interest*, 1 (4), Summer.

Mercille, J. 2011. Violent narco-cartels or US hegemony? The political economy of the 'war on drugs' in Mexico. *Third World Quarterly*, 32 (9), 1637–1654.

Minella, A. 2009. Construyendo hegemonía en América Latina, in *Los Condicionantes de la Crisis en América Latina: Inserción internacional y Modalidades de Acumulación*, edited by E. Arceo and E. Basualdo. Buenos Aires: CLACSO, 139–184.

Mining Watch Canada. 2005. *A Policy Framework for the Regulation of Canadian Mining Companies Operating Internationally* [Online]. Available at: http://www.miningwatch.ca/sites/www.miningwatch.ca/files/Policy_Framework_0.pdf [accessed: 31 May 2010].

Mining Watch Canada. 2006. *Life Before Profit! Development and Peace Kicks Off Two Year Campaign on Mining* [Online, 24 October]. Available at: http://www.miningwatch.ca/life-profit-development-and-peace-kicks-two-year-campaign-mining [accessed: 4 March 2012].

Mohan, G. and K. Stokke. 2000. Participatory development and empowerment: The dangers of localism. *Third World Quarterly*, 21 (2), 247–268.

Mollman, M. 2003. Gagging democracy. *Human Rights Dialogue*, 2 (9), June 19.

Monasterios, K. 2007. Bolivian women's organizations in the MAS era. *NACLA Report on the Americas* [Online, 12 April]. Available at: https://nacla.org/node/1460 [accessed: 15 August 2008].

Monten, Jonathan. 2005. The roots of the Bush doctrine: Power, nationalism, and democracy promotion in U.S. strategy. *International Security*, 29 (4), 112–156.

Morton, A. 2005. Change within continuity: The political economy of democratic transition in Mexico. *New Political Economy* 10 (2), 181–202.

Morton, A. 2007. *Unravelling Gramsci: Hegemony and Passive Revolution in the Global Political Economy*. London: Pluto Press.

Morton, A. 2011. *Revolution and State in Modern Mexico: The Political Economy of Uneven Development*. Lanham: Rowman and Littlefield.

Morrell, J. 2003. *A Broad Based Coalition Offers Hope: Paper Read at LASA* [Online: Haiti Democracy Project]. Available at: http://www.haitipolicy.org/content/476.htm [accessed: 31 March 2006].

Muller, E. 1995. Economic determinants of democracy. *American Sociological Review*, 60 (6), 966–982.

Muravchik, J. 1992. *Exporting Democracy: Fulfilling America's Destiny*. Washington, DC: AEI Press.

National Defence. 2011. International Operations: Crews Congratulated for Role in Counter-Drug Operation [Online, 14 November]. Available at: http://www.navy.forces.gc.ca/cms/4/4-a_eng.asp?id=880 [accessed: 1 March 2012].

National Lawyers Guild. 2004. *Summary Report of Phase II of National Lawyers Guild Delegation to Haiti* [Online, 12–19 April]. Available at: http://www.nlginternational.org/report/Haiti_delegation_report_phaseII.pdf [accessed: 1 April 2006].

Natural Resources Canada. 2009. Overview of Trends in Canadian Mineral Exploration 2007 [Online, 4 June]. Available at: http://www.nrcan-rncan.gc.ca/mms-smm/busi-indu/cme-ome/2007/cge-eng.htm [accessed: 4 August 2011].

NDI. 2005. *CEPPS/NDI Quarterly Report: April 1 to June 30, 2005, Bolivia Political Party Development* [Online: Report for USAID, June 30]. Available at: http://pdf.usaid.gov/pdf_docs/PDACF585.pdf [accessed: 3 June 2011].

NDI and Department for International Development. 2005. *Peru's Political Party System and the Promotion of Pro-Poor Reform: Synthesis Report* [Online]. Available at: http://www.accessdemocracy.org/files/1853_pe_propoor_030105_full.pdf [accessed: 14 May 2009].

NED. 2004. *Description of 2004 Grants: Latin America and the Caribbean* [Online]. Available at: Available at http://www.ned.org/grants/04programs/grants-lac04.html [accessed: 25 July 2009].

NED. 2005a. *Annual Report* [Online]. Available at: http://www.ned.org/publications [accessed: 26 July 2009].

NED. 2005b. *NED Resumes Programs in Haiti* [Online: Democracy Newsletter]. No longer available [accessed: 8 November 2006].

NED. 2006. *Annual Report* [Online]. Available at: http://www.ned.org/publications [accessed: 26 July 2009].

NED. 2007. *Annual Report* [Online]. Available at: http://www.ned.org/publications [accessed: 25 July 2009].

NED. 2008. *Annual Report* [Online]. Available at: http://www.ned.org/publications [accessed: 26 July 2009].

NED. 2009. *Annual Report* [Online]. Available at: http://www.ned.org/publications [accessed: 4 January 2012].

NED. 2010. *Annual Report* [Online]. Available at: http://www.ned.org/publications [accessed: 4 January 2012].

Neufeld, M. 2004. Pitfalls of emancipation and discourses of security: Reflections on Canada's 'security with a human face.' *International Relations* 18 (1), 109–123.

Neufeld, M. 1995. Hegemony and foreign policy analysis: The case of Canada as middle power. *Studies in Political Economy* 48, Autumn, 7–30.

Neuhouser, K. 1998. Transitions to democracy: Unpredictable elite negotiation or predictable failure to achieve class compromise? *Sociological Perspectives*, 41 (1), 67–93.

Noriega, R. 2009. A coup in Honduras. *Forbes* [Online, 29 June]. Available at: http://www.forbes.com/2009/06/29/zelaya-chavez-coup-honduras-opinions-contributors-roger-noriega.html [accessed: 8 July 2009].

North, L., V. Patroni and T. Clark. 2006. *Community Rights and Corporate Responsibility: Canadian Mining and Oil Companies in Latin America.* Toronto: Between the Lines.

Obando, E. 2006. U.S. policy toward Peru: at odds for twenty years, in *Addicted to Failure: U.S. Security Policy in Latin America and the Andean Region*, edited by B. Loveman. Lanham: Rowman & Littlefield Publishers: 169–198.

OAS. 2006. *Report of the Electoral Observation Mission in Bolivia Presidential and Prefects Election 2005* [Online, 8 May]. Available at: http://www.oas.org/sap/docs/permanent_council/2006/cp_doc_4115_06_eng.pdf [accessed: 4 May 2011].

OAS. 2009. *Informe de la Misión de Observación Electoral sobre el Referéndum Revocatorio del Mandato Popular celebrado en Bolivia el 10 de Agosto de 2008* [Online: CP/doc. 4429/09, 1 September]. Available at: http://www.oas.org/sap/docs/misiones/2008/CP%20Informe%20MOE%20Bolivia%20Agosto%202008.pdf [accessed: 4 May 2011].

Office of the Auditor General of Canada. 1996. *1996 November Report of the Auditor General of Canada.* November.

Office of the United States Trade Representative. 2011. *Americas* [Online]. Available at: http://www.ustr.gov/countries-regions/americas [accessed: 8 December 2011].

Onis, Z. and F. Senses. 2005. Rethinking the emerging post-Washington consensus. *Development and Change*, 36 (2), 263–290.

Paley, D. 2006. Minerals, gas and spin-offs: CIDA's resource regulations projects in Bolivia. *The Dominion*, 40, November.

Panitch, L. and S. Gindin. 2003. *Global Capitalism and American Empire.* London: Fernwood Publishing.

Paredes, I. 2010. Sectores exigen pruebas contra USAID y retan a Evo a expulsarla. La Razón [Online, 23 June]. Available at: http://www.la-razon.com/version.php?ArticleId=3946andEditionId=11 [accessed: 4 May 2011] .

Parish, R. and M. Peceny. 2002. Kantian liberalism and the collective defense of democracy in Latin America. *Journal of Peace Research*, 39 (2), March, 229–250.

Patzi Paco, F. 2004. *Sistema Comunal: Una Propuesta Alternativa Al Sistema Liberal. Una Discusión Teórica Para Salir De La Colonialidad y Del Liberalismo.* La Paz: Centro de Estudios Alternativos.

Paus, E. 2011. Latin America's middle income trap. *Americas Quarterly* (Winter).

Payne, A. 2000. Rethinking United States–Caribbean relations: Towards a new mode of trans-territorial governance. *Review of International Studies*, 26 (01), 69–82.

Perez-Rocha, M. 2011. Obama in Latin America: Another missed opportunity. *Foreign Policy in Focus* [Online, 24 March]. Available at: http://www.fpif.org/articles/obama_in_latin_america_another_missed_opportunity [accessed: 24 November 2011].

Persaud, R. 2001. *Counter-Hegemony and Foreign Policy: The Dialectics of Marginalized and Global Forces in Jamaica.* Albany: State University of New York Press.

Petras, J. and H. Veltmeyer. 2009. *What's Left in Latin America?: Regime Change in New Times.* Farnham: Ashgate Publishing.

Phillips, T. 2009. *South American Nations Question U.S.–Colombia Military Base Agreement* [Online: Americas Program, Center for International Policy, 14 September]. Available at: http://www.cipamericas.org/archives/1829 [accessed: 15 September 2009].

Poder Legislativo. 2004. *Ley No. 28321. El Peruano:* Lima.

Poole, D. 2010. El buen vivir: Peruvian indigenous leader Mario Palacios. *NACLA Report on the Americas* [Online, 8 September]. Available at: https://nacla.org/node/6734 [accessed: 5 May 2011].

Poole, D. and G. Renique. 2011. The Ollanta Humala victory in Peru: Moving beyond neoliberalism? *La Diaspora Peruana* [Online, 22 June]. Available at: http://peruimmigrationdocumentationproject.blogspot.com/2011/06/ollanta-humala-victory-in-peru-moving.html [accessed: 4 August 2011].

Prada Alcoreza, R. 2008. Análisis de la nueva constitución política del estado. *Crítica y Emancipación*, 1 (1), 35–50.

Pratt, C. 1990. *Middle Power Internationalism: The North South Dimension.* Kingston: McGill-Queen's University Press.

Pratt, C. 2001. Moral vision and foreign policy: The case of Canadian development assistance, in *Ethics and Security in Canadian Foreign Policy*, edited by R. Irwin. Vancouver: UBC Press, 59–76.

Presidency of the Federative Republic of Brazil. 2011. *Brazil–China, Fact Sheet* [Online]. Available at: http://www.brasil.gov.br/para/press/files/fact-sheet-brazil-china-trade [accessed: 14 September 2011].

Quixote Center. 2004. *Emergency Haiti Observation Mission* [Online, 23 March–2 April]. Available at: http://ijdh.org/pdf/PressAcounts/3.23.2005%20Emergency%20Haiti%20Observation%20Mission.pdf [accessed: 4 April 2006].

Rae, B. 2010. *Exporting Democracy: The Risks and Rewards of Pursuing a Good Idea.* Toronto: McClelland & Stewart.

Randall, S. 2010a. *Canada, the Caribbean and Latin America: Trade, Investment and Political Challenges* [Online: Canadian International Council, No. 9].

Available at: http://www.opencanada.org/wp-content/uploads/2011/05/ Canada-the-Caribbean-and-Latin-America_-Trade-Investment-and-Political- Challenges-Stephen-J.-Randall.pdf [accessed: 13 March 2012].

Randall, S. 2010b. *Canada and the Americas: Human Rights, Development and Foreign Aid* [Online: Canadian International Council, August, No.8]. Available at: http://www.opencanada.org/wp-content/uploads/2011/05/Canada-and- the-Americas_-Human-Rights-Development-and-Foreign-Aid-Stephen-J.- Randall.pdf [accessed: 14 December 2011].

Red Intercontinental de Promoción de la Economía. Social Solidaria. 2007. *Il Encuentro Latinoamericano de Economía Solidaria Y Comercio Justo: Declaración de La Habana* [Online: Red Intercontinental de Promoción de la Economía Social y Solidaria]. Available at: http://www.ripesslac.net/LDT/ RCL/05/01.pdf [accessed: 18 August 2009].

Red Participación y Justicia. 2006. *Informe Final de la Observación Electoral: Documento Final* [Online, February]. No longer available [accessed: 14 June 2009].

Reding, A. 2004. *Democracy and Human Rights in Haiti.* New York: World Policy Institute (New School University), Project for Global Democracy and Human Rights.

Remy, M. Isabel. 2005. *Los Múltiples Campos de la Participación Ciudadana en el Perú: Un Reconocimiento del Terreno y Algunas Reflexiones.* Lima: Instituto de Estudios Peruanos.

Republic of Haiti. 2007. *Growth and Poverty Reduction Strategy Paper: Making a Qualitative Leap Forward.*

Reyes, A. 2008. Falacias del neoliberalismo en el Perú. *Socialismo y Participación,* (105), 13–34.

Rieffer, B.A. and K. Mercer. 2005. US Democracy promotion: The Clinton and Bush administrations. *Global Society,* 19 (4), 385–408.

Rights and Democracy. 2002. Manhattan Minerals must Recognize the Legitimacy of the Municipal Referendum in Peru [Online, 14 August]. Available at: http:// www.dd-rd.ca/site/media/index.php?lang=enandsubsection=newsandid=504 [accessed: 11 June 2009].

Rights and Democracy. 2004. *Haiti: A Bitter Bicentennial, Report of a Mission by Rights and Democracy* [Online, January]. Available at: http://www.dd rd.ca/ site/_PDF/publications/demDev/haiti_en.pdf [accessed: 3 April 2006].

Rights and Democracy. 2009. *Annual Report: 2008–2009* [Online]. Available at: http://www.dd-rd.ca/site/_PDF/publications/annual_reports/ annualReport2008-2009.pdf [accessed: 13 November 2009].

Rights and Democracy. 2011. *Engaging a Changing World: Annual Report 2010– 11* [Online]. Available at: http://dd-rd.ca/site/_PDF/publications/annual_ reports/RAPPORT2010-2011.PDF [accessed: 21 February 2012].

Rights and Democracy and the Concertation pour Haiti. 2006. *An Open Letter Concerning the Climate of Insecurity in Haiti and the Mandate of MINUSTAH* [Online]. Available at: http://www.dd-rd.ca/site/what_we_do/

index.php?id=1619&subsection=where_we_work&subsubsection=country_ documents [accessed: 3 November 2006].

Rivera S., J. 2008. *Hacia una Nueva Constitución: Luces y Sombras del Proyecto Modificado por el Parlamento*. Cochabamba: FUNDAPPAC/Oficina Jurídica para la Mujer.

Robinson, W. 1996. *Promoting Polyarchy: Globalization, US Intervention, and Hegemony*. Cambridge: Cambridge University Press.

Robinson, W. 2004. Global crisis and Latin America. *Bulletin of Latin American Research*, 23 (2), 135–153.

Robinson, W. 2006. Promoting polyarchy in Latin America: The oxymoron of 'market democracy,' in *Latin America after Neoliberalism: Turning the Tide in the 21st century?*, edited by E. Hershberg and F. Rosen. New York: New Press, 96–119.

Robinson, W. 2008. *Latin America and Global Capitalism: A Critical Globalization Perspective*. Baltimore: Johns Hopkins University Press.

Robinson, W. 2011. Latin America's left at the crossroads. *Aljazeera* [Online, 14 September]. Available at: http://www.aljazeera.com/indepth/opinion/2011/09/2011913141540508756.html<blocked::http://www.aljazeera.com/indepth/opinion/2011/09/2011913141540508756.html [accessed: 19 December 2011].

Rock, A. 2006. *Reflections on Reforming the United Nations for the 21st Century* [Online: Government of Canada]. Available at: http://www.canadainternational.gc.ca/prmny-mponu/canada_un-canada_onu/statements-declarations/ambassadors-ambassadeurs/6840.aspx?lang=eng&view=d [accessed: 10 May 2009].

Rodriguez, N. and C. Menjívar. 2005. State terror in the U.S.-Latin American interstate regime, in *When States Kill: Latin America, the U.S., and Technologies of Terror*, edited by N. Rodriguez and C. Menjívar. Austin: University of Texas Press, 3–27.

Rojas, C. 2009. Securing the state and developing social insecurities: The securitization of citizenship in contemporary Colombia. *Third World Quarterly*, 30 (1), 227–45.

Romer, N. 2008. Interview with former Bolivian justice minister Casimira Rodríguez. *NACLA Report on the Americas* [Online, 1 September]. Available at: http://nacla.org/node/4973 [accessed: 5 December 2011].

Rousseau, S. and F. Meloche. 2002. *Gold and Land: Democratic Development at Stake*. Montreal: Rights and Democracy.

Rueschemeyer, D., E. Huber and J. Stephens. 1992. *Capitalist Development and Democracy*. Chicago: University of Chicago Press.

Russell, R. 2011. The development of inter-American relations in the past decade. *A Decade of Change: Political, Economic, and Social Developments in Western Hemisphere Affairs* [Online: Inter-American Dialogue]. Available at: http://www.thedialogue.org/page.cfm?pageID=32&pubID=2732&s=economics [accessed: 1 March 2012].

Sachs, J. 2004. The fire this time in Haiti was US-fueled. *Taipei Times*, 1 March.

Schmitz, G. 2004. The role of international democracy promotion in Canada's foreign policy. *IRPP Policy Matters*, 5 (10), November.

Schraeder, P. 2003. The state of the art in international democracy promotion: Results of a joint European–North American research network. *Democratization*, 10 (2), 21–44.

Schuller, M. 2006. *Break the Chains of Haiti's Debt* [Online: Jubilee USA Network, 20 May]. Available at: http://www.jubileeusa.org/fileadmin/user_upload/Resources/Policy_Archive/haitireport06.pdf [accessed: 4 June 2006].

Schuller, M. 2007. Seeing like a 'failed' NGO: Globalization's impacts on state and civil society in Haiti. *Political and Legal Anthropology Review*, 30 (1), 67–89.

Schuller, M. 2008. 'Haiti is finished!' Haiti's end of history meets the end of capitalism, in *Capitalizing on Catastrophe. Neoliberal Strategies in Disaster Reconstruction*, edited by N. Gunewardena and M. Schuller. Lanham: Altamira Press, 191–214.

Schuller, M. 2009. Haiti Needs New Development Approaches, Not More of the Same' [Online: Avec-Papiers, 8 June]. Available at: http://www.avec-papiers.be/Home/?p=80 [accessed: 16 July 2011].

Schulman, G. and R. Nieto. 2011. Foreign aid to mining firms: CIDA teams up with NGOs to do development work at mine sites. *The Dominion* [Online, 19 December]. Available at: http://www.dominionpaper.ca/articles/4300 [accessed: 3 March 2012].

Scipes, K. 2005. An unholy alliance: The AFL-CIO and the National Endowment for Democracy (NED) in Venezuela. *Z-Net* [Online, 10 July]. Available at: http://www.thirdworldtraveler.com/Labor/SolidarityCtr_AFL_Venez.html [accessed: 14 November 2006].

Scott, J. and K. Walters. 2000. Supporting the wave: Western political foundations and the promotion of a global democratic society. *Global Society*, 14 (2), 237–257.

Scott, J. and Steele, C. 2005. Assisting democrats or resisting dictators? The nature and impact of democracy support by the United States National Endowment for Democracy, 1990–99. *Democratization*, 12 (4), 439–460.

Seligson, M. and J. Carrión. 2007. *The Political Culture of Democracy in Peru: 2006* [Online: USAID, Latin American Public Opinion Project]. Available at: http://www.vanderbilt.edu/lapop/ab2006/peru1-en.pd [accessed: 5 July 2009].

Serrano, R. 2011. Obama defends Atty. Gen. Eric Holder amid fast and furious probe. *Los Angeles Times* (Online, 6 October). Available at: http://articles.latimes.com/2011/oct/06/news/la-pn-obama-holder-atf-20111006 [accessed: 3 December 2011].

Shamsie, Y. 2004. Building 'low-intensity' democracy in Haiti: The OAS contribution. *Third World Quarterly*, 25 (6), 1097–1115.

Shamsie, Y. 2006. It's not just Afghanistan or Darfur: Canada's peacebuilding efforts in Haiti, in *Canada Among Nations: Minorities and Priorities*, edited by

A. Cooper, D. Rowlands and D. Malone. Montreal: McGill-Queens University Press, 209–231.

Shamsie, Y. 2007. The international political economy of democracy promotion: Lessons from Haiti and Guatemala, in *Promoting Democracy in the Americas*, edited by D. Boniface, S. Lean and T. Legler. Baltimore: Johns Hopkins University Press.

Shamsie, Y. 2009. Export processing zones: The purported glimmer in Haiti's development murk. *Review of International Political Economy*, 16(4), 649–672.

Silva, E. 2009. *Challenging Neoliberalism in Latin America*. Cambridge: Cambridge University Press.

Skeen, L. 2010a. US praise for Peru's economy misses the mark. *NACLA Report on the Americas* [Online, 13 September]. Available at: https://nacla.org/node/673 [accessed: 14 March 2011].

Skeen, L. 2010b. Dubious progress in Bolivia–U.S. reconciliation. *NACLA Report on the Americas* [Online, 19 July]. Available at: http://nacla.org/node/6668 [accessed: 3 December 2011].

Sprague, J. 2007. *The Labor Battle in Haiti: Hegemony and Counter-Hegemony* [Online: Narcosphere]. Available at: http://narcosphere.narconews.com/ notebook/jeb-sprague/2005/11/the-labor-battle-haiti-hegemony-and-counter-hegemon [accessed: 12 November 2007].

Stokes, D. 2005. *America's Other War*. London: Zed Books.

Stotzky, I. 1997. *Silencing the Guns in Haiti: The Promise of Deliberative Democracy*. Chicago: University of Chicago Press.

Sum, N.L. 2008. *From 'Integral State' to 'Integral World Economic Order': Towards a Neo-Gramscian Cultural International Political Economy*. Institute for Advanced Studies in Social and Management Sciences University of Lancaster. Cultural Political Economy Working Paper Series, Working Paper No. 7.

Swan, M. 2009. Development and Peace faces new abortion-funding allegations from Peru. *The Catholic Register*, June 12.

Sweeney, C. 2009. *From Rightist Chaos to Leftist Constitutionalism: The Institutionalization of Bolivian Populism* [Online: Council on Hemispheric Affairs]. Available at: http://www.coha.org/from-rightist-chaos-to-leftist-constitutionalism-the-institutionalization-of-bolivian-populism/ [accessed: 25 November 2009]

Taft-Morales, M. 2005. *Haiti: International Assistance Strategy for the Interim Government and Congressional Concerns*. Washington: Congressional Research Service.

Taft-Morales, M. 2009. *Peru: Current Conditions and U.S. Relations* [Online: Congressional Research Service, 21 July]. Available at: http://www.dtic.mil/ cgi-bin/GetTRDoc?Location=U2&doc=GetTRDoc.pdf&AD=ADA503283 [accessed: 30 January 2012].

Tarija Libre. 2008. *USAID Contrato a Politicos Neoliberales para Conspirar Contra Evo* [Online]. Available at: http://www.tarijalibre.tarijaindustrial.

com/2008/09/usaid-contrato-a-politicos-neoliberales-para-conspirar-contra-evo/ [accessed: 7 May 2009].

Terra/AFP. 2007. *Gobierno de Chávez Acusa de Difamación a Comisión Andina de Juristas Protestors* [Online]. No longer available [accessed: 20 June 2009].

Thede, N. 2002. *Democratic Development 1990–2000: An Overview.* Montreal: International Centre for Human Rights and Democratic Development, April 1.

Thede, N. 2008. Human security, democracy and development in the Americas: the Washington consensus redux? *Canadian Journal of Latin American and Caribbean Studies*, 33 (65), 33–56.

Thede, N. and M. de la Fuente. 2008. Chronique d'un affrontement annoncé. *La Chronique des Amériques*, (8–15), September.

Thompson, G. and R. Nixon. 2009. Leader ousted, Honduras hires U.S. lobbyists. *The New York Times*, October 7.

Trouillot, M.R. 1990. *Haiti, State Against Nation: The Origins and Legacy of Duvalierism.* New York: Monthly Review Press.

UMPABOL. 2008. *Union de Mujeres Parlamentarias de Bolivia – UMPABOL* [Online]. Available at: http://www.eueombolivia.org/2006/english/ Absolutely%20Final%20Report%20Bolivia%20EOM.pdf [accessed: 15 June 2011].

UNDP. 2007. *Human Development Report 2007/2008.* New York: Palgrave Macmillan.

US Census Bureau. 2010a. *Foreign Trade Statistics Haiti* [Online]. Available at: http://www.census.gov/foreign-trade/balance/c2450.html [accessed: 12 January].

US Census Bureau. 2010b. *Foreign Trade Statistics Peru* [Online]. Available at: http://www.census.gov/foreign-trade/balance/c3330.html [accessed: 12 January].

US Census Bureau. 2010c. *Foreign Trade Statistics Bolivia* [Online]. Available at: http://www.census.gov/foreign-trade/balance/c3350.html [accessed: 12 January].

US Department of State. 2008a. *Investment Climate Statement* – Peru [Online]. Available at: http://www.state.gov/e/eeb/ifd/2008/100999.htm [accessed: 15 June 2011].

US Department of State. 2008b. *Investment Climate Statement Bolivia* [Online]. Available at: http://www.state.gov/e/eeb/rls/othr/ics/2009/117852.htm [accessed: 8 June 2011].

US Department of State. 2009. *Investment Climate Statement* – Haiti [Online, February]. Available at: http://www.state.gov/e/eeb/rls/othr/ics/2009/117840. htm [accessed: 4 December 2011].

US Department of State. 2011. *10th Anniversary of the Inter-American Democratic Charter: Remarks as Prepared for Delivery by W. Burns, Deputy Secretary* [Online: Valparaiso, Chile, 3 September]. Available online: http://www.state. gov/s/d/2011/171646.htm [accessed: 4 March 2012].

United States Senate. 2006. *Non-Governmental Organizations and Democracy Promotion, 'Giving Voice to the People': A Report to Members of the Committee on Foreign Relations.* Washington DC: 109th Congress, Second Session, December 22.

USAID. 2000. *Congressional Presentation FY 2000* [Online]. No longer available [accessed: 5 September 2006].

USAID. 2002a. *Congressional Budget Justification: Volume Two, Foreign Operations, FY 2004* [Online: Department of State]. Available at: http://www.usaid.gov/performance/cbj/index.html [accessed: 18 December 2011].

USAID. 2002b. *USAID/Peru Strategic Plan: FY 2002–2006* [Online]. Available at: http://www.usaid.gov/pe/downloads/approvedperustrategicplanfy2002-06.pdf [accessed: 3 June 2009].

USAID. 2003. *Congressional Budget Justification: Volume Two, Foreign Operations, FY 2003* [Online: Department of State]. Available at: http://www.usaid.gov/performance/cbj/index.html [accessed: 18 December 2011].

USAID. 2004. *Congressional Budget Justification: Volume Two, Foreign Operations, FY 2004* [Online: Department of State]. Available at: http://www.usaid.gov/performance/cbj/index.html [accessed: 18 December 2011].

USAID. 2005a. *At Freedom's Frontiers: A Democracy and Governance Strategic Framework* [Online, December]. Available at: http://pdf.usaid.gov/pdf_docs/PDACF999.pdf [accessed: 11 September 2006].

USAID. 2005b. *Congressional Budget Justification: Volume Two, Foreign Operations, FY 2004* [Online: Department of State]. Available at: http://www.usaid.gov/performance/cbj/index.html [accessed: 18 December 2011].

USAID. 2006a. *Congressional Budget Justification: Volume Two, Foreign Operations, FY 2006* [Online: Department of State]. Available at: http://www.usaid.gov/performance/cbj/index.html [accessed: 18 December 2011].

USAID. 2006b. *America's Accountability/ Anti-Corruption Project: Quarterly Performance Report, October – December 2005* [Online: Submitted to USAID, prepared by Casals and Associates Inc., January 26]. Available at: http://pdf.usaid.gov/pdf_docs/PDACG265.pdf [accessed: 14 April 2010].

USAID. 2006c. *USAID/ OTI Bolivia Field Report* [Online, July–September]. Available at: http://www.usaid.gov/our_work/cross-cutting_programs/transition_initiatives/country/bolivia/rpt0906.html [accessed: 5 April 2010].

USAID. 2007a. *Congressional Budget Justification: Volume Two, Foreign Operations, FY 2007* [Online: Department of State]. Available at: http://www.usaid.gov/performance/cbj/index.html [accessed: 18 December 2011].

USAID. 2007b. *USAID/ OTI Bolivia Field Report* [Online, January–March]. Available at: http://pdf.usaid.gov/pdf_docs/PDACJ863.pdf [accessed: 5 April 2010].

USAID. 2008a. *USAID Annual Performance Report: Fiscal Year 2008* [Online]. Available at: http://pdf.usaid.gov/pdf_docs/PDACM303.pdf [accessed: 19 December 2011].

USAID. 2008b. *Congressional Budget Justification: Volume Two, Foreign Operations, FY 2008* [Online: Department of State]. Available at: http://www.usaid.gov/performance/cbj/index.html [accessed: 18 December 2011].

USAID 2008c. *The Peruvian Pro-Decentralization (PRODES) Program, Final Report* [Online: USAID report prepared by ARD inc.]. Available at: http://www.bvcooperacion.pe/biblioteca/bitstream/123456789/2317/1/BVCI0002532.pdf [accessed: 18 August 2009].

USAID. 2009. *Congressional Budget Justification: Volume Two, Foreign Operations, FY 2009* [Online: Department of State]. Available at: http://www.usaid.gov/performance/cbj/index.html [accessed: 18 December 2011].

USAID. 2010. *Congressional Budget Justification: Volume Two, Foreign Operations, FY 2010* [Online: Department of State]. Available at: http://www.usaid.gov/performance/cbj/index.html [accessed: 18 December 2011].

USAID. 2011a. *Congressional Budget Justification: Volume Two, Foreign Operations, FY 2011* [Online: Department of State]. Available at: http://www.usaid.gov/performance/cbj/index.html [accessed: 18 December 2011].

USAID. 2011b. Fast Facts on the U.S. Government's Work in Haiti [Online, January]. Available at: http://www.usaid.gov/ht/docs/eqdocs/usg_factsheets/USG_FUNDING_FINAL_EDITED.pdf [accessed: 14 August 2011].

USAID-Bolivia. 2003. *Strategic Objective Close out Report: Increased Citizen Report for the Bolivian Democratic System* [Online]. Available at: http://pdf.usaid.gov/pdf_docs/PDACD699.pdf [accessed: 15 April 2009].

USAID-Haiti. 2006. *Operational Plan FY 2006* [Online]. Available at: http://pdf.usaid.gov/pdf_docs/PDACH418.pdf [accessed: 2 December 2006].

Van Cott, D.L. 2005. *From Movements to Parties in Latin America: The Evolution of Ethnic Politics.* Cambridge: Cambridge University Press.

Vanderbush, W. 2011 Good neighbor imperialism: U.S.–Latin American relations under Obama. *New Politics* [Online: 8 (3)]. Available at: http://newpol.org/taxonomy/term/122 [accessed: 12 February 2012].

Velásquez, E. and D. Antonio. 2007. EEUU armo aparato de análisis financiero para conspirar contra estabilidad económica del país. *Ciberandes*, [Online, 30 August]. Available at: http://ciberandes.co.m/index.php?id=204andtx_ttnews[tt_news]=395andtx_ttnews[backPid]=1andcHash=aeb2d3337b [accessed: 2 June 2009].

Vorbe, C. 2010. Earthquake, humanitarianism and intervention in Haiti. *Latin American Studies Association Forum*, 16 (3), 16–19.

Wade, T. 2011. Peru's Humala picks ex-army officer to lead Cabinet. *Reuters* [Online, 10 December). Available at: http://www.reuters.com/article/2011/12/10/us-peru-primeminister-idUSTRE7B90LG20111210 [accessed: 13 December 2011].

Webber, J. 2008. Bolivia: A coup in the making? *NACLA Report on the Americas* [Online, 12 September]. Available at: http://nacla.org/node/5016 [accessed: 14 May 2011].

Webber, J. 2011a. *From Rebellion to Reform in Bolivia: Class Struggle, Indigenous Liberation, and the Politics of Evo Morales*. Chicago: Haymarket Books.

Webber, J. 2011b. Fantasies aside, it's reconstituted neoliberalism in Bolivia under Morales. International Socialist Review [Online, 76 (March–April)]. Available at: http://isreview.org/issues/76/debate-bolivia.shtml [accessed: 14 May 2011].

Weidemann Associates, Inc. 2008. *Final Evaluation of USAID/Peru Poverty Reduction and Alleviation (PRA) Activity* [Online: USAID report, May]. Available at: http://pdf.usaid.gov/pdf_docs/PDACN110.pdf [accessed: 29 May 2009].

Weisbrot, M. 2006. Peru's Election: Background on Economic Issues. [Online: Center for Economic and Policy Research, April]. Available at: http://www.cepr.net/documents/peru_background_2006_04.pdf [accessed: 14 August 2011].

Weisbrot, M. 2009. *U.S. Group That Supported Overthrows of Democratically Elected Governments in Haiti and Venezuela Will Observe Elections in Honduras* [Online: Center for Economic and Policy Research, 23 November]. Available at: http://www.cepr.net/index.php/press-releases/press-releases/group-that-supported-overthrws-will-observe-honduras-elections [accessed: 2 December 2009].

Weisbrot, M. 2011. Is this MINUSTAH's 'Abu Ghraib moment' in Haiti? *The Guardian* [Online, 3 September]. Available at: http://www.guardian.co.uk/commentisfree/cifamerica/2011/sep/03/minustah-un-haiti-abuse [accessed: 4 September 2011].Welsh, J. 2004. *At Home in the World: Canada's Global Vision for the 21st Century*. Toronto: Harper Collins.

Welsh, J. 2007. Promoting democracy abroad: is it the right time for Canada to take on this file? *Literary Review of Canada* [Online, 1 December]. Available at: http://reviewcanada.ca/reviews/2007/12/01/promoting-democracy-abroad/ [accessed: 13 January 2008].

Wiarda, H. and H. Kline. 2007. The Latin American Tradition and Process of Development, in *Latin American Politics and Development*, edited by H. Wiarda and H. Kline. Boulder: Westview Press, 3–79.

Wikileaks. 2005a. Third National Cocalero Congress a Bust [Online: 05LIMA1418, 23 March Subject, US Embassy in Lima]. Available at: http://wikileaks.vicepresidencia.gob.bo/THIRD-NATIONAL-COCALERO-CONGRESS-A [accessed: 12 August 2011].

Wikileaks. 2005b. *Countering Chavez in Peru* [Online: 05LIMA4983, US Embassy in Lima, Monserrate- Margulies email, 17 September]. Available at: http://wikileaks.ch/cable/2005/11/05LIMA4983.html#par1 [accessed: 12 August 2011].

Wikileaks. 2006. *Humala Down but Still ahead in Puno* [Online: 06LIMA658, US Embassy in Lima, RR Ruehweb, February 16]. Available at: http://www.cablegatesearch.net/cable.php?id=06LIMA658&version=1305801720 [accessed: 12 August 2011].

Williams, G. 1989. Canada in the international political economy, in *The New Canadian Political Economy*, edited by W. Clement and G. Williams. Kingston: McGill-Queen's University Press, 116–137.

Williams, R. 2005. *Culture and Materialism: Selected Essays*. London: Verso.

Wilpert, G. 2003. Collision in Venezuela. *New Left Review*, 2, 101–116.

Wilson, R., B. Gills and J. Rocamora. 1993. *Low Intensity Democracy: Political Power in the New World Order*. London: Pluto Press with the Transnational Institute.

Witness for Peace. 2011. Merida Initiative 'Plan Mexico' Fact Sheet [Online]. Available at: http://www.witnessforpeace.org/downloads/Mex_Merida%20 Initiative%20factsheet%20WFP2.pdf [accessed: 1 December 2011].

Wolff, J. 2011. *Challenges to Democracy Promotion: The Case of Bolivia* [Online: Carnegie Paper, March]. Available at: http://carnegieendowment. org/2011/03/30/challenges-to-democracy-prootion-case-ofbolivia/3ha [accessed: 15 April 2011].

Wollack, K. 2005. *Challenges to Democracy in Latin America and the Caribbean. Statement made before the House International Relations Subcommittee on Western Hemisphere Affairs* [Online, 9 March]. Available at: http://www.ndi. org/files/1813_lac_testimony_030905.html [accessed: 28 May 2009].

World Bank. 2012. *Data – Indicators* [Online]. Available at: http://data.worldbank. org/indicator [accessed: 23 January 2012].

Yashar, D. 2005. *Contesting Citizenship in Latin America*. New York: Cambridge University Press.

Zibechi, R. 2009. *Massacre in the Amazon: The U.S.–Peru Free Trade Agreement Sparks a Battle over Land and Resources* [Online: Americas Program, Center for International Policy, 16 June]. Available at: http://www.cipamericas.org/ archives/1748 [accessed: 14 August 2011].

Zizek, S. 2008. Democracy versus the people. *New Statesman* [Online, 14 August]. Available at: http://www.newstatesman.com/books/2008/08/haiti-aristide-lavalas [accessed: 13 June 2009].

Index

THE INTERNATIONAL POLITICAL ECONOMY OF NEW REGIONALISMS SERIES

Other titles in the series

Our North America
Social and Political Issues beyond NAFTA
Edited by Julián Castro-Rea

Community of Insecurity
SADC's Struggle for Peace and Security
in Southern Africa
Laurie Nathan

Global and Regional Problems
Towards an Interdisciplinary Study
*Edited by Pami Aalto, Vilho Harle
and Sami Moisio*

The Ashgate Research Companion to
Regionalisms
*Edited by Timothy M. Shaw, J. Andrew Grant
and Scarlett Cornelissen*

Asymmetric Trade Negotiations
*Sanoussi Bilal, Philippe De Lombaerde
and Diana Tussie*

The Rise of the Networking Region
The Challenges of Regional Collaboration
in a Globalized World
*Edited by Harald Baldersheim, Are Vegard
Haug and Morten Øgård*

Shifting Geo-Economic Power of the Gulf
Oil, Finance and Institutions
*Edited by Matteo Legrenzi
and Bessma Momani*

Building Regions
The Regionalization of the World Order
Luk Van Langenhove

National Solutions to Trans-Border
Problems?
The Governance of Security and Risk
in a Post-NAFTA North America
Edited by Isidro Morales

The Euro in the 21st Century
Economic Crisis and Financial Uproar
María Lorca-Susino

Crafting an African Security Architecture
Addressing Regional Peace and Conflict
in the 21st Century
Edited by Hany Besada

Comparative Regional Integration
Europe and Beyond
Edited by Finn Laursen

The Rise of China
and the Capitalist World Order
Edited by Li Xing

The EU and World Regionalism
The Makability of Regions in the 21st
Century
*Edited by Philippe De Lombaerde
and Michael Schulz*

The Role of the European Union in Asia
China and India as Strategic Partners
*Edited by Bart Gaens, Juha Jokela
and Eija Limnell*

China and the Global Politics
of Regionalization
Edited by Emilian Kavalski

Clash or Cooperation of Civilizations?
Overlapping Integration and Identities
Edited by Wolfgang Zank

New Perspectives on Globalization and
Antiglobalization: Prospects for a New
World Order?
Edited by Henry Veltmeyer

Governing Regional Integration for
Development: Monitoring Experiences,
Methods and Prospects
*Edited by Philippe De Lombaerde,
Antoni Estevadeordal and Kati Suominen*

Europe-Asia Interregional Relations
A Decade of ASEM
Edited by Bart Gaens

Cruising in the Global Economy
Profits, Pleasure and Work at Sea
Christine B.N. Chin

Beyond Regionalism?
Regional Cooperation, Regionalism and
Regionalization in the Middle East
*Edited by Cilja Harders
and Matteo Legrenzi*

The EU-Russian Energy Dialogue
Europe's Future Energy Security
Edited by Pami Aalto

Regionalism, Globalisation
and International Order
Europe and Southeast Asia
Jens-Uwe Wunderlich

EU Development Policy
and Poverty Reduction
Enhancing Effectiveness
Edited by Wil Hout

An East Asian Model for Latin
American Success
The New Path
Anil Hira

European Union and New Regionalism
Regional Actors and Global Governance
in a Post-Hegemonic Era.
Second Edition
Edited by Mario Telò

Regional Integration and Poverty
*Edited by Dirk Willem te Velde
and the Overseas Development Institute*

Redefining the Pacific?
Regionalism Past, Present and Future
*Edited by Jenny Bryant-Tokalau
and Ian Frazer*

Latin America's Quest for Globalization
The Role of Spanish Firms
*Edited by Félix E. Martín
and Pablo Toral*

Exchange Rate Crises
in Developing Countries
The Political Role of the Banking Sector
Michael G. Hall

Globalization and Antiglobalization
Dynamics of Change in the New
World Order
Edited by Henry Veltmeyer

Twisting Arms and Flexing Muscles
Humanitarian Intervention and
Peacebuilding in Perspective
*Edited by Natalie Mychajlyszyn
and Timothy M. Shaw*

Asia Pacific and Human Rights
A Global Political Economy Perspective
Paul Close and David Askew

Demilitarisation and Peace-Building
in Southern Africa
Volume II – National and
Regional Experiences
*Edited by Peter Batchelor
and Kees Kingma*

Demilitarisation and Peace-Building
in Southern Africa
Volume I – Concepts and Processes
*Edited by Peter Batchelor
and Kees Kingma*

Persistent Permeability?
Regionalism, Localism, and Globalization
in the Middle East
*Edited by Bassel F. Salloukh
and Rex Brynen*

The New Political Economy of United
States-Caribbean Relations
The Apparel Industry and the Politics
of NAFTA Parity
Tony Heron

The Nordic Regions and the European Union
*Edited by Søren Dosenrode
and Henrik Halkier*